*Weighing the Future*

CRITICAL ENVIRONMENTS: NATURE, SCIENCE, AND POLITICS

*Edited by Julie Guthman and Rebecca Lave*

The Critical Environments series publishes books that explore the political forms of life and the ecologies that emerge from histories of capitalism, militarism, racism, colonialism, and more.

# Weighing the Future

RACE, SCIENCE, AND PREGNANCY TRIALS
IN THE POSTGENOMIC ERA

Natali Valdez

UNIVERSITY OF CALIFORNIA PRESS

University of California Press
Oakland, California

© 2022 by Natali Valdez

Library of Congress Cataloging-in-Publication Data

Names: Valdez, Natali, 1986– author.
Title: Weighing the future : race, science, and pregnancy trials in the
    postgenomic era / Natali Valdez.
Other titles: Critical environments (Oakland, Calif.) ; 9.
Description: Oakland, California : University of California Press, [2022] |
    Series: Critical Environments : Nature, Science and Politics ; 9 |
    Includes bibliographical references and index.
Identifiers: LCCN 2021021912 (print) | LCCN 2021021913 (ebook) |
    ISBN 9780520380134 (cloth) | ISBN 9780520380141 (paperback) |
    ISBN 9780520380158 (epub)
Subjects: LCSH: Clinical trials—Social aspects—United States. | Clinical
    trials—Social aspects—Great Britain. | Clinical trials—Political aspects
    —United States. | Clinical trials—Political aspects—Great Britain. |
    Pregnant women—United States—Case studies. | Pregnant women—
    Great Britain—Case studies. | Epigenetics—United States. |
    Epigenetics—Great Britain. | Human reproduction—Environmental
    aspects—United States. | Human reproduction--Environmental
    aspects—Great Britain. | BISAC: SOCIAL SCIENCE / Anthropology /
    General | HEALTH & FITNESS / Pregnancy & Childbirth
Classification: LCC R853.C55 V35 2022  (print) | LCC R853.C55 (ebook) |
    DDC 610.72/4—dc23
LC record available at https://lccn.loc.gov/2021021912
LC ebook record available at https://lccn.loc.gov/2021021913

30   29   28   27   26   25   24   23   22   21
10   9   8   7   6   5   4   3   2   1

*Para Natali, la niña.*
*Juntas encontramos una salida.*

# Contents

# Illustrations

# Acknowledgments

You are holding this book in your hands thanks to time, labor, care, and love. I am grateful to the people and systems that financially, intellectually, and emotionally supported me. Financial support through graduate and postgraduate fellowships (including the NSF-GRFP, Eugene-Cota Robles, Fondation Brocher, and Wenner-Gren research grants) gave me time, resources, and the opportunity to build a supportive network. And when systems failed to meet my needs, my friends, family, and mentors provided the safety net.

The PhD program in anthropology at the University of California, Irvine allowed me to explore many different projects and even supported me in taking time away from my PhD to study in the School of Public Health at UC Berkeley for a year. That experience was so valuable. It allowed me to be curious and deeply explore various aspects of my research. I am grateful to my communities in graduate school who sustained me throughout the peaks and valleys: Leksa Lee, Lydia Zacher Dixon, Véronique Fortin, Cheryl Deutsch, Georgia Hartman, Kavior Moon, Justin Perez, Daina Sanchez, and Raphaëlle Rabanes. I am also thankful to all my mentors at UC Irvine.

During the first few summers of graduate school I attended STS camp, in northern California. Not only was it a gorgeous nature retreat, but it was a fun way to connect with an extended network of scholars that supported me through my various projects, including Joe Dumit, Kate Darling, Marine Lappé, and Anna Jabloner. During my field research, I received so much support and encouragement from all of the staff members at the SmartStart and StandUp trials. Without their generosity, this project would not exist. I was supported by multiple writing groups to complete the first draft of this project in the final years of working on my PhD. Risa Cromer and Dána-Ain Davis were valuable members of the Race and Reproductivities group.

After earning my PhD, I had more time to conceptualize the book project during a two-year postdoctoral fellowship at Rice University in the Center for the Study of Women, Gender and Sexuality. The CWGS and the Department of Anthropology supported my reading groups, talks, and book workshop. I am especially grateful to Andrea Ballestero, Cymene Howe, Rosemary Hennessey, Brian Riedel, Eugenia Georges, Shannon Iverson, and Rachel Afi-Quinn. During my postdoc, I had time to create and attend meaningful conferences, including the Race and Reproduction Conference funded by Wenner-Gren that Daisy Deomampo and I organized, which brought together a wonderful group of scholars and mentors. Through the generous invitations of colleagues, I attended conferences where I was able to connect with more feminist mentors in reproduction and STS, including Charis Thompson and Rene Almeling.

I am fortunate to have had inspiring colleagues who support my work, like the entire Women of Color faculty group at Wellesley, Susan Ellison, and Eve Zimmerman. The countless walks and meditation classes that Eve and I did together nourished my soul.

My East Coast writing and support network expanded further in the last two years. I am eternally grateful to Banu Subramaniam, who has guided me through my book process and mentored me so graciously. She also connected me to brilliant feminist scholars through the Feminist STS reading group. At a workshop hosted by Jennifer Hamilton and Charlotte Kroløkke, Hannah Landecker generously read and provided vital feedback on what is now chapter 6 of this book. I also participated in a variety of queer faculty of color writing groups with Vivian Huang, Moya Bailey, and Chanda Prescod-Weinstein, all of whom have generously supported my process. The entire Nutrire CoLab was so helpful in navigating the publishing process and reviewing my writing, including Alyshia Gálvez, Megan Carney, and especially Emily Yates-Doerr (who generously read this entire book with her graduate students and gave me feedback on every chapter).

Rebecca Herzig pushed me to ask myself, at a personal level, why I was writing a book. It became clear that the lifegiving motivation for me was the relationships I built in the process. I owe a special thanks to Justin Perez for providing copyediting along with loving affirmation; Risa Cromer for seeing me in ways I wish to see myself; Leksa Lee for her loyal support and accountability; Maia G'iladi for her unfailing encouragement; and Isabel Gómez for pushing me to write every day and showing me how to find pleasure in the process. The framing of the second-person vignettes throughout the book is thanks to Isabel's intellectual generosity and brilliance. I am deeply indebted to Shannon Iverson, who was involved in the developmental editing of the entire text. You can thank Shannon for any clarity that comes through in the arguments of these chapters.

I am grateful to Kate Marshal, Enrique Ochoa-Kaup, and the entire team at UC Press for shepherding the book through review and production (during a pandemic) and for believing in the project. I am especially thankful to Dána-Ain Davis and to Rene Almeling, who reviewed the entire manuscript thoroughly and helped me in the revision process. Thank you to Vincanne Adams and the reviewers from MAQ who helped develop chapter 5.

Finally, I am everlastingly grateful to my sisters, Michele and Nicol, my mom and dad, and the extended Valdez clan. I am inspired by *mi abuelita*, the matriarch of the Valdez family, and *mi abuelito*, may he rest in peace. Dad, thanks for always having confidence in me; Michele, thank you for all your loyal support; and M&M, thank you for bringing Siënna and Mila into this world. Their births were the best rewards for meeting my writing deadlines during the last phase of a long journey.

# Introduction

You are pregnant.

You are pregnant, and as your mom explains, "heavy set"; or as your cousin says, *nalgona*; or as the tags on your jeans read, "curvy"; or as your doctor terms it "obese"; or as you say, fat.

You are pregnant, fat, and you decide to participate in a clinical trial on diet and exercise because your doctor suggested it might help you, and because someone in the waiting room of your prenatal clinic came up to you and gave you some brochures talking about how weight and diet can affect your pregnancy and your child's health. You don't want what happened last time to happen again, so you call up the number on the brochure, and the woman on the phone schedules your first clinical trial appointment.

You go to your usual prenatal clinic, and instead of meeting with your doctor or nurse, you meet a staff member from the trial. The first thing she does is tell you to "step on the scale, please." She explains that she needs to weigh you twice for accurate measurements. You don't hesitate for a second because you're used to stepping on the scale for every single medical appointment.

The woman hands you some paperwork and asks you to fill out a questionnaire. The questions are: Do you ever feel like harming yourself? Have

1

you ever binged or skipped a meal to control your weight? Do you feel less enthusiastic about your usual activities?

You cringe, and your body tightens up.

You're thinking of your last pregnancy and the loss. You keep sifting through the papers, and the woman starts talking again: "The reason we are doing this trial is to help women gain a healthy amount of weight during pregnancy. Gaining too much weight can put a woman at risk for GDM [gestational diabetes mellitus], increased weight retention, and health problems *in the future* for both mother and baby."

You don't pay close attention to the description of the trial because it sounds similar to what you've already heard.

You finish the mental health survey, but you tell the woman that the questions are weird because you're grieving. How are you supposed to answer these questions when you're grieving, pregnant, and feel tired all the time?

The woman apologizes profusely and thanks you for coming to the appointment.

Then she says, "This is an opportunity to be involved in cutting edge research that's part of a larger consortium across the whole country . . . to be involved in something big!"

For a moment you think it sounds exciting.

She goes on to describe the two groups; one is called a control or standard care, which is just like your regular appointments. The other group is more intense; you have to count your steps, count your calories, weigh yourself every day, and take meal replacement shakes.

You interrupt her and ask about the shakes. You've been so nauseous you're worried about having to drink shakes. She explains that you're given a customized meal plan based on the calories you need to limit weight gain throughout pregnancy. She explains that the US maternal health policy recommends that pregnant women with a high body mass index (BMI) should only gain about half a pound per week.

You think this sounds a bit extreme, but your doctor suggested it might help, so you stay and listen to the woman describe the "assessments": "Both groups have to complete assessments three times during pregnancy, then twice after delivery. The assessments include 7 teaspoons of blood, urine sample, blood pressure, hip, arm, thigh circumference measurements, an oral glucose tolerance test, and weight measurements."

She asks, "Have you ever done an oral glucose tolerance test for gestational diabetes?" You start to explain that in your last pregnancy you had a miscarriage at twenty-four weeks, so you never took a diabetes test.

The woman pauses in recognition, then moves on to review more information about the trial. She explains that someone from the trial will come to your house or to the hospital within seven days of your delivery to collect weight and height measurements. She says something like "unidentified biological samples like blood and urine are stored at repositories." Then she asks how you are feeling.

You feel a lot of different things. But first you tell her that you're a weird case because you had so many complications in your last pregnancy and had so much blood taken throughout. You feel traumatized, but you say that you feel really nervous about blood. You say that you need to check in with your doctor once more before deciding. You go home without signing the consent forms.

A week later, you get a call from the woman from the trial. You're busy with work, so you don't respond. After running it by your doctor one more time, who confirms that it is safe to participate in this type of clinical trial while pregnant, you decide to enroll. Another week passes and she calls again, and you decide to set up the next appointment.

You meet at the same place, and this time you sign all the papers, even the one that says you consent to having genetic tests done on your blood samples and on the cord blood samples collected at birth. But you make one final request: "Please make sure that my personal ID is not going to be associated with this trial, I don't want to be famous."

·  ·  ·  ·  ·

Why do I place you, the reader, as the "pregnant person" in this narrative? It is to make you feel, understand, and empathize with the process involved in participating in a pregnancy trial. This narrative, and book, are about the tensions between the social, political, and lived experiences of "maternal environments" and the scientific view of such environments. The maternal environment in pregnancy trials is defined as individual pregnant bodies and behaviors. With such definitions in hand, however, is the maternal environment at fault for childhood obesity or toxic chemical exposure? The key provocation of this book is that while science and society may frame pregnant people as uniquely and totally responsible for the welfare of growing fetuses and children, pregnancy and reproduction are not individual processes. *We all encompass the maternal environment.* We all collectively participate in reproduction, regardless of sex, race, gender, orientation, ability, or fertility.[1] We all contribute to the social, institutional, and environmental circumstances that shape each pregnancy, birth, and child.

How the maternal environment is operationalized in science and society is a political project; framing the maternal environment as only pregnant bodies and behaviors, rather than understanding it as everything that could influence a growing fetus including systems of poverty and racism, engenders social and material consequences for everyone, and particularly for vulnerable populations. Taking our reproductive entanglements seriously and broadening the scope of what counts as the maternal environment is an intervention in individualistic and ineffective approaches to our present and future health.

As the first ethnography of its kind, *Weighing the Future* examines the sociopolitical implications of ongoing pregnancy trials in the United States and the United Kingdom, illuminating how processes of scientific knowledge production are linked to capitalism, surveillance, racism, and environmental reproduction. The maternal environments we imagine to shape our genes, bodies, and future health are tied to race and gender, as well as to structures of inequality. I make the case that science, and how we translate and imagine it, is a reproductive project that requires anthropological and feminist vigilance. Instead of fixating on a future at risk, the book brings attention to how the present—the here and now—is at stake.

The pregnancy trials, also known as prenatal trials, that I study draw from the fields of epigenetics and developmental origins of health and disease (DOHaD) to link pregnant people's behavioral choices, like diet and exercise, with future health risks. Epigenetics, or the study of gene-environment interaction and regulation, is a field of science that examines the inheritance of changes to genetic expression without changes to the DNA sequence itself. Importantly, epigenetics has ushered in a renewed interest in "the environment," which can include the molecular "junk" surrounding DNA, sugar levels in a pregnant body, and carbon dioxide levels in the atmosphere. DOHaD is a field of study that examines how exposure during critical periods like pregnancy and early development impact health across the life span. Contemporary science frames pregnancy as a critical period of development because it encompasses multiple generations in one: the pregnant body is the first generation, the fetus is the second generation, and the reproductive cells in the fetus represent the third generation.

New research programs across epigenetics and DOHaD are increasingly characterizing postgenomic science. Postgenomics marks a shift away from gene-centered approaches to inheritance and genetic expression.[2] The postgenomic era refers to the time period following the completion of the human genome project at the turn of the twenty-first century.[3] Chapter 1 outlines the fields of postgenomics, epigenetics, and DOHaD by examining their role in pregnancy studies. Epigenetic and DOHaD logics suggest that maternal behaviors and environments in the present can impact both genetic expression *and* future health outcomes. In a postgenomic era, pregnant people are uniquely made responsible for the health risks of future generations.

The scientific and media interpretations of epigenetics and DOHaD theories individualize the health risks of future generations onto pregnant bodies alone. News articles have emerged with titles such as "Prepregnancy Diet 'Permanently Influences Baby's DNA'" and "Bad Eating Habits Start in the Womb." [4] These interpretations reflect a key aspect of the burgeoning field: pregnant bodies are at the center of epigenetic knowledge production.[5] Fetuses and reproductive bodies are not only framed as central figures in epigenetics; social scientists even claim that "epigenetics is a reproductive science."[6] Guided by a critical race and feminist lens, I organize such conceptualizations of epigenetics and reproduction into a distinct framework of postgenomic reproduction, explored further later in this chapter.

The introductory vignette draws from hundreds of prenatal trial visits that I observed, listened to recordings of, reviewed through interviews, or conducted myself. My ethnographic and feminist examination of epigenetics, pregnancy trials, and future health shows that despite significant advances in science and technology, the same interventions based on individuals—rather than on structural contexts—are funded, tested, and used to inform contemporary maternal health policy on obesity and diabetes. A key issue with such individualistic approaches to examining the maternal environment is that pregnant people should not be the only ones held responsible and accountable for present and future health risks. Denying our collective participation in reproduction and continuing to promote individual lifestyle interventions draws resources away from much needed systemic and institutional change. It further risks reproducing knowledge that is not only selectively applying and interpreting new science, but ultimately ineffective in addressing health disparities across race, or what some scholars refer to as racist disparities in health.[7] By examining how contemporary pregnancy trials draw on new science, *Weighing the Future* addresses the question: How is scientific creativity foreclosed by social and political contexts?

## PREGNANCY TRIALS IN THE UNITED STATES AND THE UNITED KINGDOM

There are currently more clinical trials that target pregnant people for lifestyle interventions than ever before.[8] Lifestyle interventions focus on changing individual bodies and behaviors and include anything from wearing a pedometer to measure steps; to using a virtual online application for diet and exercise accountability; to a variety of nutritional plans, some of which include replacing meals with shakes. In the past two decades the National Institutes of Health in the United States and United Kingdom have invested hundreds of millions of dollars in pregnancy trials aimed at understanding future health risks associated with obesity during pregnancy.[9] Pregnancy studies have a long history, and only in the past decade or so have maternal health recommendations integrated aspects of epigenetic science and DOHaD to create clinical trials that focus on how food and exercise interventions can impact future health (see chapters 1 and 2).[10]

International lifestyle pregnancy trials are similar in that they target large, ethnically/racially diverse sample sizes; focus on individual behavior changes; and are mainly funded, designed, and implemented in the Global North. I use the term *Global North* in both a critical and practical sense; it reflects how race and empire are materially and conceptually mapped onto grounded networks, resources, and methodologies of scientific knowledge production, and it captures how most, if not all, evidence-based medicine is funded and designed in North America and Europe.[11] If you were pregnant and recruited into a clinical trial in the United States or the United Kingdom, there would be some key differences, but the process of moving through each phase of the trial would be similar. One reason for this is that prenatal trials are designed in a standard way to facilitate the generalizability of data and results internationally.

The main differences that emerged as significant in my analysis across the United States and United Kingdom were contexts of health-care infrastructures and distinct histories of racialization. Most of the chapters move across the United States and United Kingdom together, but I specifically address the distinct milieus of racialization in these different countries in chapter 3. In 1993, the National Institutes of Health (NIH) in the United States required that all publicly funded clinical trials include "women and minorities."[12] This policy was also taken up in the United Kingdom. As a result, the pregnancy trials that I study in the United States and United Kingdom made a significant effort to recruit ethnically/racially diverse populations, which were distinctly defined and classified in each national context. However, despite the NIH mandate to include more diversity in clinical research, the intended impact of reducing health disparities across racialized groups has not been realized.

Throughout the book I make the case that including diverse groups of people in research and comparing health outcomes across (unstable) categories of race does not ameliorate health disparities because this approach does not effectively examine the role of racism in shaping health outcomes. Diversity and inclusion efforts do not directly address long-term exposure to unequal and unhealthy living conditions. Including more diverse people in clinical research is no doubt a worthy and necessary cause, but I argue that to address issues of equity, evidence-based medicine can and must do better. The conclusion and epilogue provide alternative ways of framing the problems and solutions to future health.

Pregnant people in the United States can access prenatal care through private insurance or Medicaid.[13] Anyone who receives Medicaid in the United States is also automatically enrolled in multiple forms of state surveillance that are not applicable to privately insured people. The United Kingdom has the National Health Service (NHS), a state-funded system that provides some free health-care options.[14] Prenatal and post-partum care is provided to all UK residents through the NHS. A unique aspect of the NHS is that midwives provide a significant portion of pre-natal care. However, who counts as a resident and what health services are available have shifted drastically in the past decade due to anti-immigrant sentiments, Brexit, and massive budget cuts to the NHS.[15]

A snapshot of the maternal and infant health outcomes across the United States and United Kingdom reflects a fairly comparable landscape. The United States and United Kingdom both have higher rates of obesity during pregnancy than other high-income nations. The United States has a higher rate of maternal and infant mortality than other wealthy coun-tries, including the United Kingdom.[16] Black women in the United States have two to three times the rate of premature birth and maternal mor-tality compared to white women. The US racial disparity in premature birth and death has not changed in the past sixty years, and some trace this disparity back further, to slavery.[17] Similarly, Black women in the United Kingdom have a much higher chance of maternal mortality and pregnancy complications than white British women, regardless of the dif-ferent health-care systems.[18]

## EPISTEMIC ENVIRONMENTS

To situate pregnancy trials and the relevant stakes within broader social and political milieus, I use the concept of *epistemic environments*, inspired by the notion of epistemic infrastructures, which Michelle Murphy defines as the ideas, methods, and economic and political structures that shape how science unfolds.[19] My employment of *environments* instead of *infra-structures* is related to the empirical material at hand: epistemic environ-ments emphasize the ways in which epigenetics and postgenomics have brought a renewed significance to the concept of environments across reproduction. "The environment" and what it includes is precisely what is at stake in the production of scientific, medical, and reproductive knowl-edge in a postgenomic era.

Epistemes refer to the ideas or logics that structure knowledge production.[20] Epistemic environments are the ideas and logics that shape contexts of scientific knowledge production. Each chapter in this book sheds light on the epistemic environments that shape the production of contemporary pregnancy trials, situated within postgenomic reproduction. I examine how racism, gender binaries, capitalism (here conceptualized as racial-surveillance biocapitalism), late liberalism (including neoliberalism), future-oriented or speculative frameworks of value and risk, and the postgenomic era collectively contribute to the epistemic environments of scientific knowledge production.[21] For example, racist hierarchies that undergird categories of ethnicity and race in scientific studies, standards that define health through individually measurable variables like weight and BMI, and heteronormative logics that define the maternal environment only as cis-gendered pregnant bodies and behaviors are all epistemes that structure the production of contemporary pregnancy trials. Ethnographically studying the implementation of contemporary pregnancy trials illuminates the epistemic environments (or milieu of ideas, and logics) that shape the imagined problems and solutions of future health.

By applying the framework of epistemic environments, my analysis of pregnancy trials makes clear that individual lifestyle interventions need to be read as symptomatic of systemic racism, rather than a solution to multidimensional illnesses like diabetes and obesity that disproportionately impact communities of color. This is because the underlying logics of individual lifestyle interventions are cut from the same ideological cloth that assumes poor, fat, and ethnically diverse individuals have risky bodies and are responsible for changing their bodies and behaviors.[22] Individual lifestyle interventions are framed as if all bodies live in similar environments and have equal access to "healthy" opportunities, choices, and material conditions. Pregnant people classified as "high risk" for diabetes and obesity are targeted for individual lifestyle interventions, while living in racist and poorly resourced environments that make it nearly impossible to comply with the intervention during the trial or to sustain the intervention changes after the trial is completed.

The epistemic environments I characterize in this book collectively shape how science imagines, manages, and apprehends future health. Such contexts and prioritization of future health have life and death consequences in the present. Despite having new knowledge about how social and political aspects of the environment—like racism—can get "under our skin" and impact health, the trials I examine selectively apply aspects of epigenetics and DOHaD research to justify individual lifestyle interventions during pregnancy. The selective interpretation of new science forecloses the maternal environment to include pregnant bodies and behaviors alone. This is one instantiation of what I call *epigenetic foreclosure*, explored later in this chapter, which illustrates the insights gained and lost by examining epigenetics using traditional evidence-based methods.[23] To understand how scientific innovation and creativity are foreclosed, my analysis focuses on systems and relations of power.[24]

*Speculative Science, Capital, and Surveillance*

In June 2014, at a conference that was funded by the NHS and European Union, the lead investigator of the UK prenatal trial that I explore in this book stated, *"pregnancy is the window into the health of our next generation."*[25] For the principal investigator (PI) of the UK trial and many other scientists at that conference, the main focus was on preventing, predicting, and speculating about future health problems. And targeting pregnancy was a means of reaching into the future.[26]

The shift toward foreseeing the future within the domains of science and health is referred to as the "speculative turn in life science."[27] Anticipatory regimes of the life sciences are shaped by capitalist ideologies that value the future over the present.[28] Twenty-first-century scientific and health-care climates are characterized by a focus on risk prevention, self-care, and individual responsibility.[29] These aspects reflect late neoliberal policies that promote the disinvestment of public health and the privatization of biological research. Predictive and personalized medicine (for those that can afford it) are manifestations of speculative health markets. Speculative and anticipatory logics undergird a wide variety of data collection practices that continue to target individual bodies and behaviors.[30] The issue with future-oriented and anticipatory risk regimes is that speculative agendas simultaneously invest scarce public funding in predictive cures, or individual interventions, and disinvest in social and public safety nets that are fundamental to population health.

A more insidious effect of future-oriented vision is that certain raced and gendered bodies are framed as risky in the present and are made responsible and subject to intervention in the name of future health. At the same time, particular reproductive futures are devalued through violent state-endorsed policies indexed by mass incarceration, high rates of premature birth and death among Black and Indigenous populations, and the destruction of families at the US-Mexico border.[31] We also see the devaluation of certain futures mapped onto the planet, people's bodies, reproductive experiences, and material conditions in day-to-day life. The valuation of profits and deregulation of the government at the expense of environmental protection unequally harms ecosystems and the bodies and lives of poor people of color.[32]

Within such conditions, how do we still imagine the solution to future health as lying solely in the hands of individuals, when the conditions and environments that impact bodies are vastly inequitable across race, class, and gender? One way to begin to address this question is to explore emergent forms of capitalism and their impact on framing health problems and solutions.[33]

Capitalisms are dynamic, and they do not unfold in the same way everywhere. Some call the United States a low-wage capitalism, characterized by declining incomes, dismantled labor unions, and widespread inequality.[34] Cedric Robinson's work on racial capitalism lays the foundation for understanding how Western capitalism was founded on the exploitation and racialization of Black bodies.[35] Black feminists such as Sylvia Wynter, Hortense Spillers, and Saidiya Hartman further theorize the fundamental role that race, gender, *and* reproduction played in the extraction and exploitation of bodies in the aftermath of slavery.[36] These literatures provide the theoretical framework for understanding how the power relations of exploitation that structured slavery remain relevant for understanding contemporary forms of exploitation in capitalist contexts across reproduction and in relation to race/racism.[37]

Another characteristic of capitalism is that it is flexible and lively and can take on new forms that arise from distinct processes of dispossession.[38] The emergent forms of capital accumulation (made possible through innovations in science and technology to extract resources, exploit labor, and create surplus) that contribute to growing global income inequality include surveillance and biocapitalism.[39] Surveillance capitalism takes human experience and translates it into behavioral data, which then accrue value in markets. The collection of massive amounts of behavioral data is used to predict consumer behavior for market purposes.[40] Behavioral future markets start with our seemingly innocuous social media inputs, extracting what we reveal in repetitive word patterns in our emails, "likes," or clicks in our phones or computer browsers; these data are used to create customized product advertisements that intend to predict what we want to buy or should buy.[41]

For example, FemTech is the digital product space that targets "women's health," encompassing apps that track menstruation, fertility, weight loss, nutrition, and breast milk production.[42] These apps collect information such as frequency of bowel movement; the odor, color, and texture of vaginal discharge; and the number of orgasms in a month. For the "user" these data can predict or chart menstruation and fertility cycles or milk production, the service provided in exchange for the user's data. However, these data are also valuable to advertising companies, making FemTech and its data a billion-dollar industry.[43] Beyond the consumer context, the emphasis on "big data" to predict future outcomes is also speculatively valuable in evidence-based medicine.

The collection of health data in a clinical trial setting is part of the same business of "big data," but on a different scale and temporality. Instead of a million bits of data collected per second, the trials I study laboriously collect biobehavioral data from thousands of pregnant bodies over the course of months, years, and up to a lifetime in longitudinal birth-cohorts (which often emerge from prenatal trials). The data collected at the pregnancy trials I study include measurable behavioral information like step counts, calories, weight measurements, sleeping patterns, and biosamples, which I call *pregnant biobits*, such as blood, urine, breast milk, and cord blood. The biological samples provide genetic data that are collected multiple times during and after pregnancy. These pregnancy data can amount to hundreds of thousands of genetic samples that can be analyzed to predict health outcomes for a future medical market (for those who can afford it) or standard intervention measures for broader "risky" groups.

Surveillance capitalism focuses on behavior and experience as data, whereas biocapitalism focuses on biological matter.[44] Feminist scholars highlight how biotechnology in current economic systems extracts material from bodies to create new kinds of markets that exchange biomatter like plasma, blood, sperm, eggs, and even surrogates.[45] Biological bits are valuable in a bioeconomy in a few different ways. The material itself, like sperm or eggs, can be bought and sold (in particular private contexts and national settings).[46] This material can also be used in scientific research.[47] The boundary between biological matter as a commodity and data is blurred because data collected now for the purposes of scientific research can also be used to produce consumer health products in the future.

Most of the literature on biocapitalism within anthropology and science and technology studies (STS) has explored sites like pharmaceutical drug development and markets, assisted reproductive technologies, stem cells, and regenerative medicine.[48] This book is distinct in that it ethnographically examines ongoing international pregnancy trials. Prenatal trials provide the bodies, data, and funding that are required to understand how behavioral modifications can shape future biological outcomes. Racial-surveillance biocapitalism and contemporary forms of liberalism are key aspects of the epistemic environments that make it possible to extract, survey, and speculate on pregnancy data.

Large-scale international prenatal trials are an understudied site for the prospecting of biobehavioral pregnancy data. By ethnographically examining pregnancy trials, I found a connection across surveillance capitalism and biocapitalism: on the surface, the differences between surveillance and biocapitalism lie in the focus on behavioral or biological data (which can also be used as surplus/speculative value). Yet in a postgenomic era, pregnancy data are deemed valuable for the prediction and surveillance of *future health*, arguably another kind of future market. Disciplining an adult to count their calories and steps is profitable in the near term, but surveying and disciplining a *pregnant* person to count their calories, steps, and blood glucose levels carries potential savings and (unequal) earnings into the future.

What is the connection between the biobehavioral data that are extracted from pregnant bodies in prenatal trials and speculative value? And who benefits? These questions are fundamental to understanding the role of epigenetics and prenatal trials in a racial-surveillance-biocapitalist context (see chapter 6). Through epigenetics, as a theory that bridges nature and nurture "divides," the biological and behavioral material that is collected from pregnant bodies carries new value and is made meaningful through clinical translation, or the translation of scientific discoveries from the laboratory to human interventions and to clinical practice.

Private companies and governments can benefit financially from pregnancy biobehavioral data. In the United Kingdom, pregnancy data, including genetic data, can be shared with private industry collaborators. In this way, private companies may use the data collected through publicly funded prenatal trials to develop biomarkers, or biological signs associated with health outcomes. Governments also envision the potential value of biomarker discoveries. Similar to the US policy, the integration of biomarkers into clinical care counts as the intellectual property of private companies, which can be patented and sold to the NHS in the United Kingdom or to private/public health-care systems abroad.[49] Potential cures for obesity and diabetes are so valuable that the speculative promise of such cures is often used to justify disinvestment in public health care in the present.[50]

Important innovations have resulted from the speculative turn in the life sciences that should not be discounted.[51] My point is that overdetermined speculative agendas, solely focused on the discovery of silver bullets or simple cures, drain limited resources *away from* social and public safety nets that are fundamental to population health. In this way, speculative, future-oriented approaches disproportionately impact vulnerable populations. Those who are in the position to speculate or bet on future remedies are not the ones who suffer the consequences of failed predictions.[52] The people in charge of investing millions of dollars of public money in clinical trials in the hope of finding a cure for obesity or diabetes will not suffer the consequences of an emaciated public health-care system. We do not need predictive biomarkers to tell us who will get sick from chronic or infectious diseases. *We already know* who will get sick: poor people of color continue to get sick at disproportionate rates. This is further evidenced by the COVID-19 global pandemic, the stakes of which I explore in the epilogue.

## POLITICS OF POSTGENOMIC REPRODUCTION

The postgenomic era is often framed as a paradigmatic shift that began around the turn of the century and is indexed by the end of the Human

Genome Project, which found that genes alone do not explain the existence of human variation and development.[53] Postgenomic reproduction is an area of research that explores how postgenomics has impacted reproduction and how reproduction has become central to innovations in science, technology, and medicine in the twenty-first century.[54] It is intentionally broad in scope to include interdisciplinary examinations of the creation, manipulation, and storage of gametes and embryos (an area of research referred to as reprogenetics in the late 1990s) *and* emerging research on epigenetics and gene-editing technologies like CRISPR.[55] Postgenomic reproduction attends to the processes and practices that lie beyond gene-centric approaches to reproduction. On a microscale it refers to how the "junk" around the DNA might also shape reproduction, and on broader scales to how seemingly unrelated aspects of "the environment" are directly linked to reproduction across generations.

By extension, I develop the *politics of postgenomic reproduction*, or postgenomic reproductive politics, as a framework that critically explores the political implications of postgenomic reproduction. It examines how new science and technology emergent in postgenomics create a distinct landscape of reproductive stakes. In what follows, I lay out how my work draws from feminist scholarship on the politics of reproduction and why it is necessary to distinguish and examine politics of twenty-first-century postgenomic reproduction.

*Legacies of Reproductive Politics*

A well-established literature within medical and reproductive anthropologies contributes to what Rayna Rapp and Faye Ginsburg termed the "politics of reproduction."[56] There are many helpful concepts for examining reproductive politics, such as the reproductive justice (RJ) framework.[57] Developed by the SisterSong collective, the RJ framework brings analytical and political attention to how racism shapes reproductive experiences.[58] It examines how mass incarceration, premature death, disinvestment in public services, and environmental justice are all reproductive issues.[59] Importantly, the RJ framework moves beyond the individualistic and liberal discourse of "choice."[60]

The intersectional theorization of reproductive politics remains vital, because biopolitical interventions were (and still are) based on the assumption that women's bodies should be disciplined, controlled, and managed.[61] As a result of a fundamental shift in nation-state formation during the seventeenth century, the systematic management of *families* became fundamental to national projects.[62] Feminist scholars highlight that discourse around the "family unit" primarily referenced essentialized framings of women's bodies and behaviors.[63] For example, the nineteenth- and twentieth-century mechanisms of surveillance and management of reproduction in the United States and United Kingdom were predominantly characterized by public health campaigns that target individual mothers and mobilize heteronormative nuclear family ideologies.[64] Reproduction was and often still is conceived of only as a "woman's" issue, and the science and medicalization of reproduction have been centered solely on "women's" bodies, which are platonically defined as cis-gendered.[65] Controlling population health through the control of reproduction was operationalized at the level of the individual through ideas of responsibility that were always deeply connected to racist ideologies.[66]

Feminist scholars applying a critical race approach have long established that the control of reproduction across race is driven by an underlying logic: Black, Brown, and Indigenous reproduction threatens a white supremacist imagination of the nation.[67] Such scholarship has examined how public health and biomedicine specifically target poor, Black, and Brown populations to control and survey reproduction.[68] In so doing, they have traced the legacies of racializing and individualizing reproductive responsibility onto pregnant bodies. Contemporary examples of forced sterilization, family separation at borders, forced adoptions, and expansive juvenile detention are just a few of the legacies of state-endorsed reproductive and family policies that express racist, nationalist, xenophobic, and homophobic/transphobic logics.[69]

Most recently, feminist scholars Martine Lappé, Robin Jeffries Hein, and Hannah Landecker have laid the groundwork for understanding the intersection of reproductive politics and environmental politics, which critically analyzes the expanding role of exposure and malleability of biology.[70] Their article ends with a call for research that rethinks reproduction across an unstable terrain of concepts, which includes *environment* as embodied, boundless, and intersectional, and *reproduction* as a process beyond the narrow focus of pregnant bodies.[71] In response to this call,

I have developed the framework of postgenomic reproductive politics and highlight here the role that epistemic environments (including racism, gender bias, speculative risk logics, and twenty-first-century liberalism and capitalism) collectively play in intensifying the stakes of reproduction. A feminist and critical race lens continues to be crucial in examining how a postgenomic age can both deepen *and* ameliorate inequalities in reproductive health.

Guided by Black feminism, postcolonial science and technology studies, and the anthropology of reproduction, the politics of postgenomic reproduction critically attends to issues of race and gender in two main ways: its application is based on the premise that processes of racism are enacted in and through reproduction, and that queering reproduction requires a reframing of the maternal environment that is not biologically and genetically essentialized or individualized to cis-gendered, reproductively capable bodies.

*Twenty-First-Century Politics of Postgenomic Reproduction:*
*A Feminist Tool and Framework*

The purpose of conceiving the politics of postgenomic reproduction is to delineate the layering of older biopolitics and reproductive politics within postgenomic reproduction. Conscripting pregnant people into believing that they have complete control and responsibility over fetal development and epigenetic modifications is an example of this prevailing approach to reproduction. As an overarching framework, postgenomic reproductive politics examines the science, technology, processes, methods, and practices that make it possible to extract resources like pregnancy biobehavioral data for the creation of future health markets and surveillance. It thus serves as a feminist tool for examining early twenty-first-century scientific knowledge production, translation, and application of postgenomics as a reproductive project.

I situate my research on pregnancy trials within postgenomic reproduction, but not entirely or cleanly. Pregnancy trials simultaneously draw on technologies, methodologies, and biopolitical strategies that *precede* a postgenomic age, and they translate scientific theories from *within* a postgenomic era. As a hybrid phenomenon, contemporary pregnancy trials spotlight the durable ideas and approaches of reproductive and biopolitical projects from the twentieth and twenty-first centuries. In this way, contemporary pregnancy trials that draw on epigenetics lie at the nexus of knowledge that is imagined as both old and new, that has formed in dissonant, uneven, layered manners. For instance, chapters 1 and 2 provide historical and political analyses of how, despite the substantial shifts in science and technology in the twenty-first century, lifestyle interventions applied in contemporary pregnancy trials have not fundamentally changed since the 1950s. In addition, twenty-first-century applications of pregnancy trials are not novel in how they make individuals (predominantly poor and people of color) responsible and vulnerable to state intervention and surveillance in the name of population health.

Yet what emerges as distinctly indexical of the politics of postgenomic reproduction are the ways in which different forms of capitalism (specifically, a constellation of racial-surveillance biocapitalism) and liberalism converge at the site of nascent pregnancy trials and how this convergence promotes the accumulation of biobehavioral data for future markets. Moreover, by ethnographically examining contemporary pregnancy trials, I found that such applications of evidence-based medicine represent a unique mobilization of epigenetics as a reproductive science; they reflect the expansion of reproductive surveillance and management across the life course and into future generations. Put another way, examining epigenetics via contemporary pregnancy trials exemplifies the politics of postgenomic reproduction.

For example, the politics of postgenomic reproduction as a conceptual framework reveals distinct forms of the optimization of life beyond one person, to include multiple and future generations.[72] An analysis of the politics of postgenomic reproduction examines how interventions, forms of surveillance, and biobehavioral data prospecting are operationalized at new boundaries of time and scale (gene, individual, population). Together, DOHaD and epigenetics advance the biosocial conceptual framework to apprehend future health by surveying and prospecting pregnant bodies. Pregnant bodies in particular are profoundly contested sites precisely because they are framed as "windows into future health." Moreover, the politics of postgenomic reproduction remains relevant on the micro- and macroscales of what to eat, how to eat, and when to eat, as well as on the scale of health across the life course. Through postgenomic research and the speculative turn in the life sciences, these mundane aspects of everyday life are now framed as consequential for future health.

Another emergent characteristic of postgenomic reproduction is the temporality or *chronicity of surveillance* via prenatal and postnatal care and evidence-based medicine. Now the focus is not just on gestation or early development, but also on preconception, indexed in the expansion of preconception trials. These trials target women's diet, exercise, and drug and alcohol consumption *before* pregnancy.[73] These interventions are not as concerned with current reproductive health issues like unstable housing, limited childcare, unemployment, immigration, and racism. Preconception trials reflect the prioritization of intervening in the future, which draws resources and attention away from systemic change in the present. Such trials also maintain a heteronormative focus on the reproductive capacities of cis-gendered individual bodies. What results is a process of reproductive surveillance that is never ending, across reproductive life spans and across generations. Additionally, the massive amounts of data collected on sleep, eating, and exercise behaviors in pregnancy trials have the potential to be used for the production of health surveillance tools and systems. Far less, if any, interventions target men's reproductive capacities before, during, or after conception. However, unlike reproductive politics that characterized the twentieth century, the politics of postgenomic reproduction in the twenty-first century shows more interest in expanding surveillance, attention, and research on cis-men's reproductive capacities.[74]

I examine the politics of postgenomic reproduction in relation to epigenetics, DOHaD, and pregnancy trials; however, this feminist framework remains relevant to other reproductive phenomena. For instance, as an analytical framework, it is salient to the reevaluation of the maternal environment in surrogacy, the implications of epigenetics as a reproductive science, the employment of epigenetic logics in the evolving fields of reprogenetics and reproductive technologies, and the experimentation with and implementation of gene-editing technologies like CRISPR.[75]

*Futures Imagined and Epigenetic Foreclosure*

By examining the politics of postgenomic reproduction of contemporary pregnancy trials, I found that certain applications and interpretations of epigenetics became possible and not others. I call this process *epigenetic foreclosure*. Epigenetics itself is a capacious theory, one that destabilizes discrete notions of the environment and the malleability of gene expression. In its initial integration into public realms it was regarded as a "hopeful" theory full of potential, yet its application in human trials has yielded insignificant or inconclusive results.[76]

Epigenetic foreclosure captures the unfolding of epigenetics in ambiguous ways without predetermining an outcome. For instance, a capacious understanding of the maternal environment might consider how diverse bodies, communities, and policies all contribute to reproduction.[77] Yet chapter 5 illustrates a form of epigenetic foreclosure by showing how pregnancy trials define the maternal environment only as individual pregnant bodies and behaviors. Chapter 6 reveals another valence of epigenetic foreclosure indexed by the selective interpretation of epigenetic theories. For instance, indeterminate and latent processes of epigenetic modifications are selectively ignored in pregnancy trials. By illuminating processes of foreclosure, I aim to trace how particular interpretations materialize without analyzing these emergent threads through moralized or politicized binary frameworks of good/bad, right/left.

By revealing how certain interpretations become flattened, and also emphasizing the indeterminacy around such reductive interpretations, epigenetic foreclosure aims to make visible both the closures and apertures. At the edges of foreclosure there are possibilities for how epigenetic interpretations and applications can be otherwise. Processes of epigenetic foreclosure are also iterative and multiple. Diverse areas of epigenetic science do not experience the same kinds of foreclosure. Epigenetics in molecular and neurosciences and animal and plant studies all unfold in different ways with different political implications. In addition, different translations of epigenetics into policy, research design, and standards maintain diverse forms of foreclosure.

*Weighing the Future* illustrates one valence of epigenetic foreclosure in pregnancy trials, the implications of which are woven into its narrative arc. I trace processes of epigenetic foreclosure that I encountered in US and UK clinical trials to underscore areas of scientific growth and limitation. Though postgenomic reproduction has expanded the scope and stakes of "environmental" exposure, it also reflects the limits of our own social values and scientific creativity. In the process of translating new scientific ideas into pregnancy trials, certain potentials are reduced and flattened to fit within existing social, economic, and scientific methodologies and ideologies.[78]

This book engages the broad fields of epigenetic science, reproduction, and chronic illness, but the stakes are relevant beyond these topics. Thus far ethical guidelines have constrained research such as human genome editing, but if these are circumvented, as world patterns clearly show, the geneticization of life will enter hyperdrive.[79] Already we see scientific trends center on epigenetic science and CRISPR technology; in the next ten years another theory or technology will emerge, yet we risk recreating the same approaches to health if we do not critically reevaluate the epistemic environments that undergird or contextualize our scientific systems and our imagined futures.

Paying attention to imagined scientific futures is important because science and technology reflect back to us what is deemed valuable and which bodies are worth saving and investing in, in the present. Pregnancy trials provide a site to explore how particular futures are imagined, valued, and translated into scientific knowledge. When tracing the stakes of postgenomic reproductive politics, epigenetic foreclosure, and the role of epistemic environments in constraining scientific creativity, a core question emerges: What kinds of anti-racist and gender-inclusive environments promote (not just the reproduction of but) the care and sustainability of healthy, safe beings and relations?

## BEING AND KNOWING: SITUATING RACE AND RACISM

Situating my approach to race and racism involves a degree of self-contextualization.[80] *Writing* about race/racism makes me uneasy.[81] Talking, feeling, experiencing—these realms are accessible to me; they shape my embodiment of race/racism. No word or sentence that I write does the embodied experience of race/racism justice. But others have shown me it is possible to write about race, racism, and power relations in ways that can resonate across different experiences.[82] Even with the muses and mentors, writing is not my preferred tool: perhaps because I fumbled my way through two languages growing up and erased one entirely to show my loyalty to the other; perhaps because I was told in graduate school that I was a weak writer and would not succeed as a scholar. Maybe it is because writing is not in my blood.

I am a child of Mexican immigrants. My grandparents never received a formal education. My grandfather came to the United States as a *bracero*, a laborer granted access to work in agriculture, picking seasonal fruits and vegetables. During his yearlong stints in the United States, his only way to communicate with my grandmother in Mexico was to pay someone to write letters to her. She paid someone to read them and write letters in return. Only recently in my family history have reading and writing become tools for articulating, apprehending, and situating power relations. My mother and my father were the first in their families to go to college. Now I write a book—a book framed by and through the relations of power that structure the production and access to knowledge. In this book, which would never have been written by my Brown, queer hands if the powers that be had tighter control of knowledge production, I write about race/racism.[83]

My approach to race/racism is unruly, undisciplined, and rooted in what Gloria Anzaldúa describes as a restless consciousness made to sustain contradictions.[84] The development of this is related to my own family's trajectory from (un)educated to hyper-educated. I draw from the fields of anthropology, sociology, feminist postcolonial technoscience, Black feminism, critical race studies, and queer studies to offer a feminist take on epigenetics and future health.[85] Enduring racist hierarchies within political and biological domains have organized bodies and flesh into what Silvia Wynter termed genres of humanness.[86] Such hierarchical systems of organization and classification have shaped medical and scientific processes of researching, counting, diagnosing, treating, and intervening in certain raced and gendered bodies.[87] It is with this knowledge that I have set out to analyze power relations of race and gender in postgenomic reproduction.[88]

A point of reference that influences my analysis and framing is Katherine McKittrick's theorization of race/class/gender/sexuality as *not merely indicators of identity*, but as spatial knowledge and experience.[89] Inspired by McKittrick's theorization of race and space, I frame racism in relation to epistemic environments, characterized previously, and use the term *racist environments.* Racist environments maintain systems of power that create the conditions in which certain bodies are made sick by long-term, repetitive exposure to racism. We live in a racist environment: But why does this bear repeating? I suggest that as scientific paradigms shift, so do the conceptualizations, manifestations, and obfuscations of race/racism.

Shifts in scientific paradigms throw into relief distinct framings of race/racism that require critical feminist analysis. I argue that scientific innovation is thwarted by focusing only on diversity and inclusion as a process of recruiting participants with different pigmented flesh, often linked to ever-changing ethnicity classifications. Such approaches do not systematically study how structural racism indexes and perpetuates disparities in health across race. My analysis of racist environments, derived from the concept of epistemic environments, focuses on how relations of power structure the design and implementation of the clinical trials. Put another way, I focus less on how many people of color are included in science; instead my analysis spotlights *how the obfuscation of racism* is enveloped into scientific knowledge production.

As I discuss in chapter 1, epigenetic science shows us that multiple environments exist and intersect on different scales. "The environment" in epigenetics, further explored in chapter 5, includes cells, bodies, neighborhoods, foods, climates, and experiences. While the concept of racist environments is connected to environmental racism, it stands apart because racist environments integrate the ideological structures that undergird the material and discursive manifestations of racism. Racist environments encompass the toxic pesticides and lead that pervade Black, Indigenous, and other communities of color, *as well as* the experiences of racism in medicine, science, and technology, conceived as obstetric racism, medical racism, and race as technology; the systems/structures that simultaneously target "risky" racialized groups in public health campaigns; and the racism in public/medical discourses that blame Black Americans and Latinx communities for having disproportionate rates of comorbidities (like obesity and diabetes).[90] Racist environments in a postgenomic era provide the conceptual umbrella to envision multiple and entangled forms of race/racism.

While focusing entirely on identity-based categories of race is fundamental to inclusion and diversity strategies in medicine and science, examining racist environments shifts the focus toward material and discursive systems and milieus that directly impact health outcomes. Through epigenetics, we can understand how experiences of racism can physiologically impact blood pressure, heart rate, and cortisol levels, and we can understand how racism increases individual exposure to toxic chemicals that perpetuate health impacts across generations.[91] By extension the notion of racist environments frames exposure to racism as discursive, experiential, material, and visceral. Whereas the notion of racist environments serves as a framework that can hold multiple contexts, intentions, impacts, and consequences of racism, racial improvisation is a praxis, a conceptually based methodological approach to empirically studying dynamics of racist environments. Racial improvisation examines a phenomenon, like race/racism, that can shapeshift at different rates and becomes relationally reconfigured across micro- and macroscales.[92] I explain the method of racial improvisation further in chapter 1, where I discuss my methodologies in ethnographically examining clinical trials, and apply the method in chapter 3.

In examining and theorizing processes of knowledge production in postgenomic reproduction, I am invested in illuminating the underlying relations, or epistemes, of race/racism that are veiled and obscured. For instance, whiteness, coded as normative, default, neutral, and objective, remains so through the obfuscation of enduring power dynamics. Guided by a Black feminist tradition that made marginalized, unrecognized knowledge matter in academic institutions through the theorization of experience, I aim to make visible the whiteness and racism that is maintained through denial and unrecognition. *I spotlight the contours of race/racism that lie in a negative space.* In my research on the implementation of clinical trials, scientific processes that are classified as neutral and objective actively erase processes of racialization. In chapter 3 I show how race emerges as significant for targeting and recruiting ethnically/racially diverse participants and then disappears (as discussed in chapter 6) when the biological samples extracted from diverse bodies are transformed into pregnant biobits and data. The data are analyzed as neutral objects, disconnected from the racist environments that encompass the pregnant bodies.

In mirroring the obfuscation of race/racism that occurs in processes of scientific and medical knowledge production, my analysis of race/racism surfaces to the center of some chapters and retreats to the background in others. The analysis of race/racism is thus woven across different processes and practices; some chapters spotlight forms of race/racism and others highlight its obfuscation. This approach is intentional; it shows how relations of race/racism are strategically seen and unseen.[93] If you lose the thread, return to this section. Hierarchies of power are always there; they are in the ether, atmosphere, limited only by our capacity, politics, and methods to sense and perceive them. Tuning in to the environments and improvisations of race/racism requires knowledge beyond reading and writing, including touching, feeling, smelling, hearing, and tasting embodied racism.[94] Such awareness becomes important for reimagining current health problems and future solutions.

A final note for readers: the book is organized into three parts. Chapters 1 and 2 set the stage for understanding the integration of epigenetics and DOHaD research into contemporary pregnancy trials. Both chapters also illuminate the durability of individually based pregnancy interventions despite inconclusive results. Chapters 3 and 4 explore the processes and politics of trial recruitment and intervention implementation.[95] These chapters explore my firsthand experience with recruitment and center the narratives of pregnant participants. Chapters 5 and 6 examine the methodological, racial, and scientific consequences of defining the maternal environment in pregnancy trials. In so doing, they shed light on the limits of using the randomized clinical trial (RCT) method in translating epigenetic ideas, as well as on the speculative value that is created from prospecting pregnant biobehavioral data. Together, chapters 3 through 6 ethnographically ground the selective interpretations of new science in pregnancy trials. The book concludes with the afterbirth of foreclosure, which examines the challenges and possibilities of redistributing resources and labor toward present health as a form of social justice.

# Part I

# 1   Epistemic Environments

REPRODUCING SOLUTIONS TO PAST, PRESENT,
AND FUTURE MATERNAL HEALTH

The Motherwell study, named after the town it was located in, Mother-well, Scotland, was designed by Dr. Grieve, the local obstetrician. Between 1952 and 1976, Dr. Grieve implemented a unique nutritional intervention with thousands of women, from the main hospital of the town.[1] Dr. Grieve's intervention included a calorie-controlled, low-carbohydrate, and high-protein diet (primarily red meat). The intervention prohibited women from eating bread, potatoes, prunes, plums, bananas, canned fruit, nuts, or dates, and they were not allowed to smoke, unlike other women at the time. Dr. Grieve also gave all the pregnant women strict weight gain limitations and warned that excessive weight gain would cause preeclampsia, which was consistent with the medical advice given at the time.[2]

More importantly, all the women in the study were required to eat one pound of red meat each day.[3] Dr. Grieve designed this "Atkins-type" diet because he had an interest in body building and believed that this approach during pregnancy would promote fetal growth.[4] In a booklet that provided instructions for the diet, Dr. Grieve wrote: "Quantity (of meat) is more important than quality. As it may be difficult to eat enough meat at meal times, the use of cooked meat, especially corned beef, rather than fruits or biscuits, is advised to assuage hunger between meals."[5]

The Motherwell study was implemented for over twenty years, and it generated large amounts of data and helped recruit many women and their children into subsequent follow-up studies.[6] The findings were also influential in the creation of global health policies related to maternal nutrition and weight. For instance, the study informed the maternal nutrition and weight guidelines in the United States and United Kingdom in the 1970s, and the follow-up studies were influential in shaping epigenetic and developmental origins of health and disease (DOHaD) claims about the impacts of maternal nutrition across the life course.

In the 1970s, the results of the Motherwell study found that women who received Dr. Grieve's intervention had smaller babies than women in neighboring towns. Following these results, scientific consensus further developed around the idea that restricting maternal weight did not improve infant health. One other finding from the study that did not receive much attention was that infants born within the Motherwell study had low birth weights and *lower rates of mortality* compared to those born to women in other parts of Europe and Scandinavia who experienced restricted maternal weight gain due to war and famine. This finding was counter to the existing claims that causally connected low birth weight and infant mortality.

That is, despite the low birth weights among the infants in Dr. Grieve's study, they still had a higher chance of living than infants born during famine and war who also had low birth weights. Might these contradictory differences in birthweight and mortality point to the role of the political and ecological environment that infants are born into, not only maternal weight and diet? Instead of examining how health risks linked to low birth weight and maternal weight are complexly related to environmental conditions during pregnancy, the interpretations of the Motherwell study focused entirely on creating health policies that strictly defined women's weight during pregnancy.

The Motherwell study offers an entry point for assessing the social, political, and scientific landscapes of prenatal nutrition interventions across the past and present. It represents an older kind of prenatal nutrition experiment, prior to the standardization and ethical implementation of randomized clinical trials (RCTs) with pregnant women. Nevertheless, core aspects of the Motherwell intervention remain similar to current "cutting-edge" prenatal lifestyle interventions. Despite the significant technological and ethical changes in pregnancy studies since the 1950s and a paradigmatic shift in science that often characterizes the postgenomic era, the content of the nutritional intervention itself and the premise that maternal weight and diet should be intervened in prevails as the dominant approach to maternal health care.

Yet to date, the hundreds of completed pregnancy trials that have implemented lifestyle interventions in diet and exercise have not found any conclusive evidence that prenatal nutritional interventions reduce pregnancy complications or improve infant health outcomes related to diabetes or obesity.[7] Based on the current results of prenatal nutrition trials, it is not possible to make causal, linear associations between maternal nutrition and weight only and metabolic syndromes in future generations. Focusing on maternal diet alone ignores the material and bodily effects of other intersecting aspects of the environment. Despite this poor record, the growth and investment in lifestyle interventions in diet and exercise during pregnancy remains impressively high.

How is it that current science cannot imagine other ways to interpret and evaluate existing data on maternal nutrition and health? The Motherwell findings pointed to existing environmental differences that distinguished health outcomes across infants born in different social and political milieus. These data existed in the late 1970s. More importantly, how is it that scientific communities centered in the Global North continued to design, fund, and implement nutritional interventions in 2015 that were similar to the dietary interventions of 1955? What enduring aspects from the mid-twentieth century continue to shape the current epistemic environments, such that imagined solutions to maternal and future health remain limited?

In what follows, I contextualize relevant scientific theories and methodologies such as postgenomics/epigenetics, DOHaD, and the RCT method, and how they intersect at the site of pregnancy trials.[8] In so doing, I outline the scientific and technological shifts that have taken place since the Motherwell study. Then I analyze these scientific fields, theories, and methods from a feminist and critical race perspective. In doing so, I illuminate the contours of past and present epistemic environments by interweaving politics of race, gender, and power relations. Finally, I expand on my own approaches to ethnographically studying pregnancy trials. Overall, this and the next chapter reflect the durability of individualized lifestyle interventions during pregnancy.

## SCIENTIFIC FIELDS, DEFINITIONS, AND METHODS

### Postgenomics

The postgenomic era is often used as an umbrella term to characterize the period following the sequencing of the human genome, beginning around the early 2000s. Although there is no clear consensus on whether postgenomics is a new scientific paradigm, the "post" prefix indexes a shift in understanding the significance of genes and gene-environment interactions. Prior to postgenomics, the gene-centric approach to genetics and genomics dominated the scientific imaginary of the last half of the twentieth century. The gene was understood to carry all the necessary information for organisms to develop and function. At the peak of the genetic paradigm, the Human Genome Project (HGP) sequenced the entire human genome.[9] The HGP was an endeavor full of promise and expectations, but once the project was completed in the late 1990s, scientists and the general public were surprised by the results. The HGP concluded that humans have far fewer genes than originally estimated, and only a fraction of the genome consists of genes that are used to make proteins.[10] That is, genes alone could not explain human variability or biological development. Other materials and mechanisms were at play in shaping how parts of the genome are regulated and expressed. The disenchantment with genetics as the blueprint of life opened up a space for the development of postgenomics and a newly enlivened interest in epigenetics as a field that does not center genetic determinism.[11]

Postgenomic science includes the methodological, conceptual, and technical innovations in sequencing genomes and the expansion of biomarker development and analysis.[12] The root term *genomics* is complicated to define because it is continuously unfolding and changing. A capacious understanding of genomics includes genomes (all the genes that provide instructions and information for protein synthesis) *and* the material in and around the genome. *Biomarkers* are biological signs or indicators connected to particular health outcomes and are analyzed in blood, urine, or tissue samples. They become important for tracing epigenetic modifications across maternal and infant environments, an aspect I return to in chapter 6. Importantly, the fields of epigenetics and DOHaD have significantly changed the postgenomic landscape, particularly in the realms of maternal health and reproductive sciences.

## Epigenetics

*Epigenetics* is a key field within postgenomics. The Greek prefix "epi" literally means "on top" or "above"; broadly defined, epigenetics refers to the study of gene-environment interaction and genetic expression.[13] In current understandings, epigenetics can include a variety of mechanisms involved in genetic programming, such as DNA methylation, histone modification, and noncoding RNAs, as well as the regulation of chromatin structure.[14] For instance, the environment around the DNA can include methyl groups that attach to parts of the DNA; this process is called methylation, and it can impact the instructions for an organism's development. Previously, scientists did not consider these processes to be important parts of genetic expression; however, epigenetic science spotlights these processes as important to the modification of genes and their expression.

The definition of epigenetics has changed over time. When Conrad Waddington, a British scientist, first coined the term epigenetics in the mid-twentieth century, it was derived from the term *epigenesis*, or the process of development through differentiation.[15] Waddington drew from the notion of acquired characteristics, or genetic assimilation, to illuminate epigenetic processes.[16] He explained that genetic assimilation occurs as a result of complex genetic interactions during development, and that these processes are *flexible*. For instance, Waddington claimed that an organism could assimilate or rather "remember" or embody environmental stress that occurred in past generations.[17]

The notion of genetic assimilation or acquired characteristics is similar to a much older concept of biological development credited to Jean-Baptiste Lamarck, a French naturalist (1744–1829). In the early nineteenth century Lamarck proposed a theory of inheritance through acquired characteristics (a giraffe acquires a long neck in response to surrounding tall trees, which were a main food source). By the late 1800s Lamarck's work was overshadowed by the dominant theory of evolution developed by British collaborators Alfred Wallace and Charles Darwin.[18] The notion of Darwinian evolution was supplemented by Gregor Mendel's theory of genetic inheritance, published in the early 1900s. Mendel's work explained that genes from "mother" and "father" are inherited by the offspring.[19] Whereas Lamarck's work is referenced as "soft inheritance" because it denotes flexibility and malleability of inheritance in relation to the environment, Mendel's work became known as "hard inheritance," characteristic of genetic determinism.[20]

Epigenetics in humans fundamentally challenges traditional notions of inheritance. Whereas genetics is defined by Mendelian inheritance or the idea that discrete pairs of genes are passed from parent to child only, epigenetics is defined as non-Mendelian inheritance, with the potential of both transgenerational and intergenerational inheritance. In epigenetics, environmental factors can influence genetic expression and regulation, and epigenetic modifications from past generations can be carried on to shape development across generations. As Moshe Szyf explains, "it was generally believed that most of the epigenetic information is erased during early gestation, and if this erasure were complete then errors in the epigenetic markings would not be transferred across generations."[21] He goes on to say that due to unpredictable modifications, or stochasticism, on parts of the DNA, it is possible that these changes can actually be inherited across generations and be latently triggered later in life.

Latency and stochasticism are indeterminate and flexible aspects of epigenetics that make it challenging to design, execute, and trace the impacts of exposures across fetal, infant, and child development. Often these indeterminate aspects of epigenetics are ignored in designing clinical trials.[22] Some scientists emphasize that the unpredictability and latency of epigenetic inheritance reflects gene expression potential, in that it is *not* determinable what kinds of epigenetic modifications are passed down and whether or not they are expressed.[23] Framing heritability through the concept of "potential" highlights the malleability of biological and genetic development, which challenges overdetermined ideas of genetic predisposition.

Understanding this fundamental shift in human inheritance is important for contextualizing the heighted attention to pregnancy trials that implement lifestyle interventions. Through epigenetics, fetal development is vulnerable to environmental modifications in utero. That is, fetal genes and their expression are not fixed. They can be modified during pregnancy by multiple forms of environmental exposures. Scientists refer to this as "fetal programming," or how fetal genomes can be shaped, influenced, and modified during gestation through environmental stimulation. Epigenetics has amplified the stakes of environmental exposure especially with regard to fetal programming in utero; however, definitions of environmental exposure are not stable. Within epigenetics, the environment includes the social, material, political, and experiential. In current epigenetic literature, "the environment" encompasses that which is both inside and outside the body. Importantly, though, epigenetic theories do not necessarily imply a division between inside and outside; rather, epigenetics emphasizes that the environment occurs on different *scales*, from the cellular level all the way to the atmospheric level.[24]

Another key theme of epigenetic science is the molecularization of the environment. By interviewing scientists who work on gene-environment research, Katherine Darling and colleagues found that scientists conceived of the environment as "anything non-genetic," highly molecular, and personalized.[25] The authors also found that the environment was often conflated across "internal"/"external" boundaries between bodies and environments.[26] Within environmental health agencies, the molecularization of the environment shifts focus on exposures at molecular levels instead of atmospheric toxic pollution.[27] In Alzheimer's research, the prioritization of the molecular scale shifts resources and attention away from present care and instead focuses on finding a biomarker that can predict the manifestation of Alzheimer's disease, an illustration of reductive interpretations of epigenetics.[28] The focus on the molecular scale of the environment, through biomarkers, can obscure cultural or personal experience and context.[29]

Finally, epigenetics as a biological mechanism and a field of science is not explored or understood in the same way across different disciplines. For instance, epigenetics has long been studied in plants and some animals.[30] However, in humans it is much harder to trace epigenetic modifications because human environments are not as easily controllable in a study design. For some experts, epigenetic mechanisms include primarily the molecular scale of biological processes. Environmental epigenetics—which some refer to as a distinct subfield—focuses on "how changes in the social and material environment have a physiological impact on individuals and on forms of sociality."[31] Other experts make the case that epigenetics is a "reproductive science" because tracing epigenetic mechanisms in humans across generations relies on studies that focus on reproduction, pregnancy, and early child development.[32]

In the context of prenatal trials, environmental epigenetics is relevant in the broader sense; however, no one involved in the clinical trials I examine in this book specifically used the term environmental epigenetics, instead using the general term epigenetics. The fact that different terms are employed to talk about similar areas of research regarding pregnancy trials—such as environmental epigenetics, behavioral epigenetics, nutritional epigenetics, and molecular epigenetics—further indexes the emergent unfolding of epigenetics itself.

## Developmental Origins of Health and Disease (DOHaD)

Epigenetic science is often discussed in relation to DOHaD because it provides a biological explanation for how environments can impact health long term.[33] Epigenetic mechanisms, like methylation, enable an understanding of how the "outside" gets "inside" and how this matters for health outcomes across the life span. This mechanistic link is important because DOHaD is dedicated to examining how exposure to adverse environments during early development (including pregnancy) can impact health across the life span. One kind of "exposure" that is examined is maternal nutrition; DOHaD claims that diet and nutrition during pregnancy can "predict risk" for metabolic syndromes like heart disease, diabetes, obesity, and hypertension in the next generation.[34]

Initially, DOHaD was conceived of as a "society" that came about from the meetings of the World Congress on Fetal Origins of Adult Disease, which met in 2001 and 2003 in India and the United Kingdom.[35] Since then it has expanded into a large field of study that has significantly impacted reproductive medicine and specifically the design and implementation of behavioral prenatal trials.

One person who emerges as a key figure in the development of DOHaD is David Barker, a British scientist who studied the connection between nutrition during gestation and rates of cardiovascular disease later in life.[36] He developed the "fetal origins hypothesis," a theory that is integral to DOHaD research because it focuses on how early development, specifically gestation, affects health across the life course.[37] Barker proposed that fetal programming during pregnancy can *permanently* impact the metabolic development of the offspring.[38]

To develop this theory, Barker examined medical records of children and grandchildren born to families who had experienced famines. For example, he reviewed multigenerational health records from the Netherlands (during and after the Dutch Hunger Winter) and from a northern area of Sweden that experienced periods of famine.[39] Barker and colleagues then designed the Hertfordshire cohort study, which looked at men who were born from 1911 to 1930. This group of men were then traced to examine their health outcomes in relation to their birth weights. The cohort study provided data for examining research in gene-environment interactions and disease across the life course.[40] The results of these studies found that birth weight was complexly connected to health outcomes later in life. In fact, more recent analyses, as of 2005, found that "the *postnatal environment* may be more influential [in] the development of obesity in later life."[41] I emphasize postnatal environment, because in slight contrast to this, most contemporary trials focus on the prenatal environment.

Regardless of the fact that the original DOHaD studies on famine and nutritional exposure included men and explored paternal lines, subsequent studies that Barker and his team led focused primarily on the role that individual pregnant bodies and behaviors play in fetal development, over and beyond any other environmental factors. This narrow focus is also connected to larger arguments on the molecularization of life. In the late 1990s and early 2000s, Barker and his team designed studies on the diets of pregnant women during gestation by surveying thousands of women in Southampton, United Kingdom. The findings from these studies promoted the claim that maternal diet and weight impacted infant adiposity and metabolic risks of cardiovascular disease and obesity later in life.[42] As a result, the claim that weight and diet during pregnancy shape future health for children and adults became entrenched in contemporary prenatal trial designs, and it remains one of the main justifications for the pregnancy trials that I study in this book.

*The Randomized Controlled Trial (RCT) Method*

The earlier research that shaped DOHaD theories relied primarily on observational cohort studies that retrospectively reviewed records and surveys to connect nutritional exposures in past generations to current health outcomes. These types of study designs and methodologies produce associations, connections, and relations, but not causality. Unlike observational studies, which can only reveal associations, RCTs are used to make causal associations. By the early 2000s, pregnancy studies that integrated theories of epigenetics and DOHaD were also applying the RCT method to design pregnancy studies with behavioral or lifestyle interventions.

The RCT method is currently referred to as the "gold standard" of evidence-based medicine, and it is the main method in biomedicine for testing and developing health policies. Research development in the fields of obesity and diabetes, education, economics, and global health depends primarily on the implementation of RCTs.[43] As a result of a global overdependence on the RCT method to fund health interventions and create new policies, RCTs with large sample sizes and advanced statistical power are necessary to prove that lifestyle interventions during pregnancy can impact health across the life course.

However, not all RCTs are the same. Implementing pharmaceutical or behavioral RCTs involves distinct risks and ethics. For instance, phase I drug or pharmaceutical trials only target healthy nonpregnant adult subjects and are the riskiest phase, because they aim to show that the drug is safe. Phase II trials involve proving the drug or intervention is effective on a small scale and reveal whether the drug is worth investing in on a larger scale. Phase III trials require hundreds or thousands of participants and are usually deemed safer for implementation on larger populations.[44]

The pregnancy trials that I study in this book are deemed safe for pregnant populations primarily because they are testing behavioral lifestyle interventions and not drugs. Not only are they deemed safer and less risky than drug trials, lifestyle pregnancy trials are distinctly framed as providing benefits to improve pregnancy health outcomes.[45] Despite differences in risk associated with different types of trials, the designs of behavioral or drug RCTs are very similar. In discussions around behavioral interventions, scientists focus on standardizing the "dose" of the intervention *as if it were a drug.* Behavioral prenatal trials are most similar to phase III clinical pharmaceutical trials. Both require large (diverse) sample sizes, use comparative groups and randomization, and are considered safer than phase I pharmaceutical trials.

The production and implementation of RCTs are very expensive, and they are primarily designed and funded in the Global North, which creates an uneven distribution of knowledge production and global health policy influence.[46] The disproportionate production of RCTs in the United States and Europe has to do with the sociopolitical context of when and where the method was designed and standardized. Experimental studies completed in colonial, militaristic, and resource-poor settings in the twentieth century contributed valuable insights to the current derivation of the RCT method. Methodological approaches such as quarantine periods, exposure dosages, and the development of controlled experimental arms of interventions were tested in colonial laboratories in an effort to treat local diseases (which were primarily considered a "problem" because they were affecting military occupants). For instance, hygiene campaigns in the Caribbean and Central America, as well as military sites in the Philippines, were used as experimental settings to test the effectiveness of interventions.[47] These sites were framed "as laborator[ies] for discovering and testing the elements of a global health system for the twentieth century," which included the development of methods like the RCT.[48] Once these methods were better developed, they were then translated to national domestic contexts in the Global North.[49] These historical and political contexts of colonialism matter for how contemporary scientific methods are mobilized and applied in uneven global landscapes.

Throughout the twentieth century, racial minorities and prisoners of war and of the state were subject to unethical and violent treatment in the name of scientific experimentation that applied aspects of the RCT method.[50] In response to human rights violations, the post–World War II context significantly shaped the creation of ethical guidelines in the implementation of RCTs. Not only do these histories of racism and exploitation influence how the RCT method itself was developed, they also influence, as I explain further in chapter 3, how minority populations are recruited into current pregnancy trials. The movement to institutionalize ethical requirements also impacted the participation of pregnant populations.

In particular, the "thalidomide trials" of the mid-twentieth century prompted strong restrictions on the recruitment of pregnant people into drug trials.[51] The ingestion of thalidomide during pregnancy affected fetal development, which caused hundreds of thousands of children to be born with malformed limbs. Consequently, pregnant populations became classified as a vulnerable population and are rarely involved in trials for pharmaceutical drug development due to unknown health risks to the fetus.[52] Only recently have policies shifted to allow pharmaceutical trials to target pregnant participants for drug experimentation.[53]

In the second half of the twentieth century, a pivotal phase in the development of the RCT method was its application in pharmaceutical research. Melinda Cooper argues that the expansion of mass drug markets, national health care, and biomedicine could not be possible without the standardization and institutionalization of the RCT method. With the growth of industrial production, she argues, it became "feasible, for the first time, to manage large volumes of clinical data, and to produce, measure and predict events on the scale of whole populations."[54]

Additionally, the applications of the RCT method go beyond drug markets. In the United Kingdom, the post–World War II environment shaped the standardization, implementation, and regulation of RCT methods within the National Health Service (NHS) and public welfare systems.[55] The mid-twentieth century was known as the "golden age of evaluation" in the United Kingdom.[56] Money was poured into using the RCT method to evaluate the effectiveness of interventions in complex social settings. For instance, studies used the RCT method to assess the effectiveness of sexual education in regard to health outcomes, or the impact of work incentives on employment rates. However, the application of the RCT to assess public policy interventions proved problematic because the studies consistently concluded negative findings, found minimal effects, or were inconclusive.[57]

Consequently, debates on the overdependence on RCTs to evaluate social processes that are "multi-variate and non-linear, characteristic of the real, social world" ensued in the late 1970s (primarily in North America and the United Kingdom).[58] The critiques in the 1970s claimed that RCTs proved challenging to apply to the 'real world' because such settings "demand techniques for 'controlling complexity,' and because [the real world goes] beyond simple 'if-then' causal connections."[59] Importantly, scholars in the field of education returned to these critiques of the RCT method in the early 2000s. And I argue that these same criticisms of the limits of the RCT method for capturing complex, multivariate contexts are also relevant to the field of environmental epigenetics and DOHaD with regard to contemporary pregnancy trials.

## Contemporary Pregnancy Trials

What do contemporary pregnancy trials look like in practice? Whereas the rest of the book explores this question, here I provide a brief introduction to the US and UK pregnancy trials that I explore.[60] In the United States, I examined the trial that I call "SmartStart," which was nested within a larger national consortium that included seven different RCTs in distinct locations across the United States and Puerto Rico.[61] The main outcome of the consortium that included the SmartStart trial focused on how life-style interventions targeting diet and physical activity could impact gestational weight gain; the second outcome had to do with whether these interventions during pregnancy could affect maternal and infant outcomes. The intervention that the SmartStart trial implemented included similar components to the Motherwell study. Pregnant participants had to limit their gestational weight gain by weighing themselves each day, counting calories, following a specific meal plan and exercising each day. The SmartStart trial also used meal replacement shakes, a nutritional intervention aimed at weight control.

In a published description of the consortium, the authors write that "maternal obesity during pregnancy increases health risks for both mother and child, including complications during gestation and delivery as well as *future* obesity, diabetes, and cardiovascular risk."[62] In a webinar presentation that I attended on the US trial, the justification for the trial was presented by drawing directly from epigenetics and DOHaD, which link nutrition and weight during pregnancy to health across the life course. In the webinar, maternal obesity and excess gestational weight gain were framed as part of a "vicious cycle" of obesity encompassing mother and infant. Such frameworks shape the design and justification of contemporary nutritional pregnancy trials. The current working hypothesis is that maternal obesity directly impacts the risk of obesity in the child, yet *no pregnancy trial to date* has been successful in affirming this hypothesis. In fact, an enormous number of pregnancy trials that test lifestyle interventions are not clinically significant and do not find significant effects on infant health outcomes.

While I was working on the US SmartStart trial, there were four other similar national consortiums. Altogether the five US consortiums included about twenty-eight individual RCT sites across the United States and represented hundreds of millions of dollars in funding.[63] All of these consortiums implemented individual lifestyle interventions targeting obesity during prenatal, postnatal, and early childhood periods. In order to find any clinical significance, RCTs need large sample sizes. Therefore, these national consortiums also worked together to pool their data.

In the United Kingdom, I examined the trial that I call "StandUp," which was also part of a consortium that included five sites for data collection and implementation across England and Scotland. Similar to the US SmartStart trial, the StandUp trial in the United Kingdom drew from epigenetics and DOHaD theories. The principal investigator (PI) of the UK trial referenced Barker's work in particular to explain the theories that shaped her approach to designing her pregnancy trial. The StandUp trial focused on glycemic control and not gestational weight control. Therefore, the UK intervention did not count or control calories, but rather focused on making "healthy swaps" from high glycemic foods to low ones. It also incorporated a physical activity component, and participants were able to meet with the interventionists or health trainers about eight times during their pregnancy. In chapters 3 and 4 I further describe the organizational structure and intervention of the US and UK trials.

## POLITICS OF SCIENTIFIC KNOWLEDGE PRODUCTION

Over the past half decade since the Motherwell study, the paradigmatic shift of the postgenomic era, the dissemination of DOHaD theories into maternal health, and the industrialization of RCT method have all changed the landscape of pregnancy studies. Collectively these changes shape aspects of postgenomic reproduction. Yet the design and implementation of current pregnancy studies have not fundamentally changed from the pregnancy studies that were completed in the mid-twentieth century. To be sure, how these interventions are implemented and analyzed has significantly changed, but the content and values that undergird current lifestyle interventions are familiar to older interventions.[64] The pregnancy trials that I examine in this book apply intervention components similar to those in the Motherwell study: a strict calorie-controlled meal plan, daily weight measurements, restricted weight gain, daily exercise, and specific diets. The focus remains squarely on weight control and individual lifestyle changes.

The durability of such approaches is connected to the politics of race, gender, and power.[65] For instance, critical feminist approaches to science and technology studies explain that racist and heteropatriarchical structures of dominant cultures have material impacts on scientific knowledge production.[66] That is, how science is funded, designed, implemented, and disseminated is directly related to ideologies and relations of power.[67] For instance, power relations inherent to whiteness and processes of othering are crucial to the maintenance of scientific authority.[68]

In my approach to understanding power relations inherent to postgenomic reproduction, whiteness emerged as a concept and not just a racialized category. As a concept, whiteness is also intrinsically tied to racial oppression, white supremacy projects, and social control.[69] As such, whiteness is a relation of power that shapes scientific knowledge production.[70] For instance, in the area of physics, Chanda Prescod-Weinstein characterizes *white empiricism* as the fields, systems, and disciplines dominated by ideas that are produced primarily by and for white people. Prescod-Weinstein writes, "white empiricism comes to dominate empirical discourses in physics because whiteness powerfully shapes the predominate arbiters of who is a valid observer of physical and social phenomena."[71] Similarly, whiteness as a relation of power, not merely identity, also shapes processes and practices of producing knowledge across epigenetics, DOHaD, and pregnancy trials.

Questions such as who designs the majority of clinical trials and how or why the majority of global health policies are developed in the Global North are connected to the concept of whiteness as a dominant relation of power. Most if not all pregnancy trials that test lifestyle interventions are designed by groups in North America, the United Kingdom, and Europe. Similarly, the key figures in characterizing the fields of epigenetics and

DOHaD are also based in the Global North. In a critical sense, as I mentioned in the introduction, the dominant design and funding of RCTs from within the Global North reflects how race and empire are mapped onto grounded networks, resources, and methodologies of scientific knowledge production. Whiteness as a form of social power is made evident by the conditions and structures of knowledge production, such that the same groups of people have access to elite spaces of knowledge production, and the English language is the imperial vehicle for knowledge dissemination.

The historical context of modern science from the eighteenth and nineteenth centuries also matters for how race and whiteness are mobilized in postgenomics.[72] The preoccupation with creating order in nature through scientific reason came about within the Enlightenment, and race science was one aspect within this broader project. Organizing classifications of race was part of a systematic scientific project, which also shaped the epistemic environments of Darwin's theory of evolution and subsequent theories of eugenics (promoted by Francis Galton, Darwin's cousin). Eugenic logic, a key mechanism of white supremacy, was enveloped into modern economics, public health, and medicine.[73] The emergence of modern science is explicitly tied to racist science that was employed for white supremacist ends of maintaining hierarchies of race, slavery, and oppression. Denying this context facilitates the perpetuation of racism and the power of whiteness in science. Embracing this context does not take away from what we can learn about modern science, but rather adds much needed value to present and future deployments of such knowledge.

Similar to the genetic paradigm, popular media have framed epigenetics as hopeful and full of promise, so powerful that the emergence of the postgenomic era was accompanied by debates on how race would become obsolete in this new paradigm.[74] Yet two decades into the so-called paradigm shift, the COVID-19 pandemic, along with the national and international protests against racial injustice, reveal that not only does race still matter, but *racism is still killing people.*[75] Far from being a paradigmatic shift in understanding race in science, as Nadia Abu El-Haj explains, the postgenomic era reinscribes notions of race and logics of neoliberalism that make classifications of race meaningful for blaming individuals.[76]

Furthermore, contemporary studies that purport to examine race and class in relation to epigenetics also index a form of epigenetic foreclosure.[77] Such epigenetic studies are framed by questions such as "Do the poor have different patterns of methylation than the rich?"[78] Or, are African Americans destined to higher rates of diabetes, obesity, and premature death because of transgenerational inheritance of trauma? Such interrogations use epigenetics to biologically essentialize the predisposition of health outcomes due to race and class. Recent explorations of epigenetics and race continue to focus on racial difference, not on how processes of racism impact health outcomes.[79] By failing to recognize the comprehensive role that multiple and intersecting environments play in shaping individual vulnerability to epigenetic modifications, such studies foreclose notions of the environment in epigenetics. Moreover, studies that overdetermine impacts of race and class selectively ignore the flexibility and unpredictability of epigenetic modifications; this is a form of epigenetic foreclosure that I explore in the rest of the book.[80]

## Politics of Applying and Disseminating DOHaD

The children of the Motherwell experiment, introduced at the beginning of this chapter, were enrolled in subsequent studies led by the students of David Barker, a key figure in the field of DOHaD. By tracing the health outcomes of the children born into the Motherwell study, scientists found that a high-protein diet during pregnancy increased rates of high blood pressure in the offspring.[81] It turns out that eating one pound of meat every day during pregnancy is not the "healthiest" approach to maternal nutrition.

There are few, if any, articles or commentaries discussing the ethical issues of the Motherwell study and the long-term negative health impacts of Dr. Grieve's prenatal nutrition intervention. The Motherwell study signals the ways in which science—and in this case, ideas about health, nutrition, and pregnancy—are based on culturally dominant beliefs of what is "healthy." Dr. Grieve created an intervention based on his "body-building" nutrition approach, and thousands of patients/participants followed the intervention because he was the obstetrician, the medical authority. He did not seek any external ethical review from the town or hospital.

His position of authority facilitated his intervention in women's bodies and diets for over twenty years. The compliance on behalf of the pregnant women is viscerally marked not just on the pregnant bodies, but on their children's bodies, who experienced higher blood pressure, connected observationally to their mother's excessive red meat consumption.

From my review of the publications that follow up on the Motherwell study, the authors obfuscate the connection to Dr. Grieve's intervention, and instead of naming him, authors refer to Dr. Grieve in a detached manner: "The consultant obstetrician in Motherwell encouraged expectant women to eat 1 lb of red meat per day"; "He discouraged carbohydrate-rich food"; "He recommended"; "He advised"; and "He warned [. . .] against excessive weight gain."[82] By not explicitly connecting how current studies draw from past studies that are completed in contexts of uneven power dynamics, fraught gendered/raced conceptualizations of bodies and health are continuously enveloped into contemporary theories.

Feminist theorists have recognized that power dynamics are inherent to processes of knowledge production, and bringing these issues to light does not undermine science.[83] Even if exposing uneven power dynamics does not completely resolve the inequalities or disparities directly, it is nonetheless a step in the direction of imagining alternative futures. What's more, denying or obscuring the unequal power dynamics that contextualize the production of data, which are used and valued for generations, protects the inherent power relations of whiteness in the name of scientific neutrality and objectivity. The point is that naming the contours of power and forms of racism in science facilitates the creation of health interventions that directly acknowledge a racist past, address a racist present, and avoid reproducing a racist future.[84]

Another example of the politics that undergird the dissemination and interpretation of DOHaD is the development of theories that explain the main causes of "lifestyle" diseases like obesity and diabetes. Collaborators with Barker from Southampton University went on to develop one such theory, called the "mismatch" theory, and it explains the etiology of obesity through environmental and evolutionary approaches.[85] One description of this work states: "Our bodies evolved to allow our ancestors the best chance of survival as hunter-gatherers in the Savannah. Our brains, on the other hand, have allowed us to develop complex societies, cultures, and lifestyles, far removed from those of our ancestors. As a result, write Peter Gluckman and Mark Hanson in *Mismatch*, we have created a modern artificial world that is painfully out of tune with our evolved bodies."[86] This illustration of Gluckman and Hanson's work is an example of how intersectional politics of race, gender, and power are obscured in the dissemination of scientific theories related to DOHaD. The inexact description of "our brains," which create cultures and lifestyles that hurt our bodies, produces a Wizard of Oz illusion that obscures the realities behind the curtain; racism, capitalism, neoliberalism, heteropatriarchy, and colonialism are the epistemic environments that shape the material conditions of obesity and diabetes, not merely "our brains."[87]

Whereas this summary points to broader evolutionary forces that shape obesity and diabetes, the content in *Mismatch* draws specifically from research that focuses on maternal weight and diet.[88] In the research developed to support the book's main argument, Gluckman and Hanson write that "the environment of the fetus [which] is created by its mother" influences fetal development of and health risks to the offspring.[89] This description of "the environment" further embeds the framework that a fetal environment is an individual pregnant body, automatically assumed to be the "mother." Such understandings of obesity etiology also reinforce the motivation to blame, target, and intervene in pregnant bodies as the main protagonists in shaping environmental exposures that can cause obesity.

Furthermore, the current etiologies for multidimensional chronic illnesses like obesity and diabetes are shockingly repetitive and familiar: being obese or overweight is a result of not living a "healthy" lifestyle. In a postgenomic era, the message remains the same: be disciplined, work hard, eat right, and exercise—*especially if you are pregnant*. The added caveat that epigenetics and DOHaD science have introduced in this dominant framing is that if you are fat or diabetic as a child or adult, it may be your mother's fault for being overweight during pregnancy.

Twenty-first-century capitalism (via racial-surveillance biocapitalism) and late liberalism (via policies and practices of neoliberalism) are in fact what drives the underlying notion that individuals can shape their biology, health, education, and economic standing. The myth is that anyone could be upwardly mobile (and healthier) if only they were more disciplined and took better care of themselves, all while maximizing their labor. Still, there are unspoken conditions to this myth. In order to scale current social, economic, and health inequalities, created by racist environments, the successful individual should be white, able-bodied, thin, heteronormative, and have citizenship in the Global North.

## Politics of Postgenomic Reproduction

Pregnancy trials are important to postgenomic reproduction, which includes the fields of epigenetics and DOHaD, because pregnancy trials became the ideal setting for confirming new ideas of inheritance and health across the life span. More specifically, pregnant bodies and the pregnant data excavated during trial implementation (explored in chapter 5) provide the raw materials for examining how biobehavioral modifications during gestation can impact future generations. Even though earlier studies in DOHaD and epigenetics focused on the role of paternal lines, the focus on pregnant bodies dominates the fields.[90]

The focus on pregnancy trials and pregnant bodies in epigenetics and DOHaD is often criticized as a gender bias in the design of postgenomic reproduction research.[91] However, this is not an isolated case of gender bias or negligence in reproductive science writ large. The dearth of information is well documented in the book *GUYnecology: The Missing Science of Men's Reproductive Health*, by Rene Almeling, which shows how dominant efforts in reproductive medicine over the last century have focused disproportionately on women's bodies, but questions like how men's health matters for miscarriages and childhood illness have received little to no legitimate scientific attention.[92]

In pregnancy trials, what counts as the maternal environment is a political concern with material consequences for pregnant people *and* for everyone who participates in reproduction (including those who are and are not recognized as having reproductive capacities).[93] In chapter 4, I show how the clinical translation of epigenetics defines the maternal environment as compartmentalized units like the uterine environment, metabolic environment, or fetal environment. The term *fetal environment* foregrounds the fetus, and the pregnant body stands in as the space that belongs to the fetus, which emphasizes the coalescing power around fetal rights over maternal rights.[94] Moreover, framing the maternal environment as a "fetal environment" only also risks mistreating surrogates as merely vessels and not rights-bearing people. Daisy Deomampo writes that enduring colonial power dynamics in a global capitalist setting make Indian bodies available for the implantation and gestation of embryos from wealthy, white, American and European couples.[95] Further, framing the maternal environment as only the bodies and behaviors of cisgendered women alienates and excludes trans communities, gay families, and extended kin networks of care that lie beyond the heteronormative imagination of biological, nuclear family structures.[96]

The exclusive and narrowed focus on pregnant bodies as a critical period of epigenetic modifications—and not on broader environments of all people regardless of biological reproductive abilities—is characteristic of the pervading gendered and racist politics of reproduction. These politics are relevant in the twentieth century and are newly enlivened through the politics of postgenomic reproduction in the twenty-first century. Additionally, the temporal foreclosure that occurs when only focusing on gestation is based on the speculation that this moment matters beyond other moments, a conjecture that is already being undermined by the emergent rise of preconception trials.[97]

Pregnant people are targeted for lifestyle interventions in diet and exercise, as if their individual self-control and responsibility to "eat right" could change genetic expression in their children and future generations. However, a comprehensive reading of epigenetics illustrates that pregnant individuals are not in control of epigenetic modifications (see chapter 4).[98] One of the issues with reductive or individualistic interpretations of epigenetics and DOHaD is that they ignore stochasticism or unpredictable latent manifestations and rely completely on notions of linear causality.[99]

By spotlighting areas of epigenetic foreclosure in postgenomic reproduction, my aim is to also bring attention to how current scientific understanding of epigenetics and DOHaD has only scratched the surface of the complexity inherent to the ways that behaviors, genes, and environments codetermine health outcomes. As I explain in chapter 5, the selective interpretations and applications of epigenetics and DOHaD are connected to the limits inherent to the RCT method. RCTs are dependent on the investigation of discrete environmental variables, which makes it challenging to capture the capacious scope of epigenetics in human models. The balanced takeaway is that epigenetic foreclosure is not predetermined; there are other ways of exploring the capaciousness of epigenetics. A critical and transparent examination of race, gender, and power is vital for understanding the limits and possibilities of epigenetic foreclosure.

## ETHNOGRAPHICALLY STUDYING INTERNATIONAL PREGNANCY TRIALS

This is the first book-length project that ethnographically captures the implementation of ongoing clinical trials with pregnant participants in two national settings. Gaining access and institutional permission took over three years, which required ethical approval from three separate review committees in two countries. To better understand the clinical context, approach, and methodologies of PIs and patients, I took two years of coursework in the fields of epidemiology and child-maternal health.

Although the people I worked with at my field sites are public scholars and the information about the trials is also public, I make a concerted effort to protect their identities by changing all names associated with the trials and not disclosing authors' names in citations.[100] This project is not about individuals; consequently, I do not include a lot of "thick description" about individual PIs, but I situate them within processes, practices, and networks of science. The emphasis is on the epistemic environments in which PIs are working. I had access to interview all staff members, PIs, and research collaborators associated with each trial. Since the pregnant participants were enrolled in an ongoing trial, I was not allowed to ask them my own interview questions. However, I gained information about pregnant people's experiences through my participation in and observation of the intervention phase, or the phase in which participants in the experimental group receive the lifestyle intervention, as well as my work as an interventionist.

Only in the US trial did I have dual roles. While I was collecting my own ethnographic data, I also worked as a staff member on the SmartStart trial. I was trained as an interventionist, and I worked closely with the intervention staff and PI. In the US trial, the interventionists are in charge of delivering the intervention to the pregnant participants in the intervention or experimental group of the trial. In order to observe and learn about how the trial was implemented, the PI of the SmartStart trial gave me the option to work as an unpaid staff member. As an interventionist, I met with pregnant participants in the intervention group every two weeks for about five to six months of pregnancy. We also had weekly meetings with the PI to troubleshoot any intervention issues. These sessions gave me insight into the experiences and challenges of other interventionists and their assigned participants. Tracing the experiences of the women in the intervention during their pregnancy gave me an intimate view of their everyday routines, challenges, and social conditions. Illustrations of the US trial intervention are featured in chapter 5, and analysis of the UK intervention is centered in chapter 4.

In addition to participant observation and interviews across the US and UK trials, I attended national conferences on obesity and webinar conferences on pregnancy trials, and completed archival research on histories and health policies of relevant topics. Through my preliminary research I was introduced to the problem of "consensus" regarding obesity during pregnancy. The key difference in approaching obesity during pregnancy emanated from the distinct policy recommendations: the United States focuses on gestational weight gain (GWG) recommendations, while the United Kingdom does not; I examine the details surrounding these different approaches in the next chapter. The UK policy claims that there is not enough evidence to sustain a focus on GWG. The scientists that I worked with in the United States did not understand why the scientists in the United Kingdom were adamantly against GWG and calorie-controlled interventions. One collaborator in the SmartStart trial felt so passionate about the topic that she flew to England to discuss the matter with the PI conducting the trial in the United Kingdom. Prior to going to England, she had assumed that the United Kingdom applied the same approach to obesity during pregnancy, because both the United States and United Kingdom drew from similar scientific data and publications. I followed the fault line between the two, and doing so led me to the two trials that I examine here.

The incorporation of both trials in my research is not aimed at making analytical comparisons across the sites.[101] Instead, I frame my fieldwork as a multisited ethnography, which emphasizes the connections across different spaces, networks, and actors.[102] Multisited projects explore the relationality between and among networks, without reifying geographic and national boundaries.[103] The incorporation of both the StandUp and SmartStart trials came about through conversations with key informants. The collaborators and PIs at each trial knew of each other tangentially. While I was in the field, another trial funded by the European Union approached both of the PIs at the StandUp trial and SmartStart trial to participate in a larger consortium of trials collecting behavioral data on pregnant women. These PIs met each other at the European data consortium. As I mentioned previously, the networks and trials involved in prenatal interventions are small and mainly located in the Global North. Even though the United States and United Kingdom maintain distinct national health-care contexts, the methods and trial designs in each respective place are similar, and they both intend to generalize the results of their trials so that they are applicable to diverse global populations.

## Methodological Attunements: Racial Improvisation

I developed the praxis of racial improvisation to study how racist dynamics unfold in spaces that are seemingly devoid of explicit racism. It was a method developed out of a need to ethnographically capture unpredictable moments, gestures, silences, processes, and practices that index race/racism through its implicit obfuscation and sometimes explicit reference.[104] In elite, predominantly white, spaces of scientific knowledge production, like well-funded clinical trials in the Global North, race/racism is recognized as politically incorrect and is thus intentionally unseen. Racial improvisation as an approach carries significance beyond the clinical trial setting. Racial improvisation can examine the dynamic mercuriality of racialization that occurs in processes of classification such as national census projects, job applications, and algorithms. Such processes require forms of medical standards and applications of "race" as an identity category, not necessarily as something that is experienced.

My methodological attention to improvisation is based on the following well-established premises: racism exists; racism is dynamic, creative, lively, iterative, and relational; and race and racism are forms of power relations that are emergent, unstable, and mercurial. Extant scholarship shows how such power relations are made evident through empirical examination. My intervention of improvisation emphasizes the indeterminate and unpredictable aspects of race/racism. I do not assume I know the particular kinds of categories, relations, forms, or representations that can index race/racism. Improvisation requires an open, intuitive attention.[105]

In practice, an example of the improvisation of race, as I show in chapter 3, occurs during recruitment when participants are asked about their racial/ethnic identity. Across the United States and United Kingdom, participants are provided distinct choices, and when these options do not fit the individual's identity or experience of race, they are required to improvise their race/ethnicity into an existing category. The conditions make such improvisations possible and necessary. Another example is the invocation of racist assumptions that emerge in casual chatter, such as a brief statement I heard in my fieldwork: "African men like their women heavier." This was a brief statement, unsubstantiated and emergent. Catching this requires an improvisational attention and attunement to the ways in which race/racism can arise seemingly out of nowhere. Through improvisation, my analysis does not center the individual, but rather the conditions of possibility that make these utterances viable in elite spaces of scientific knowledge production. In attending to racial improvisation, a key question is: What collective sentiments and *processes* of anti-blackness and racism make such emergent moments visible and possible?

·   ·   ·   ·   ·

By providing scientific and political context on relevant theories and methods, this chapter illustrated how the fields of epigenetics and DOHaD come together at the nexus of lifestyle pregnancy trials. The next chapter further examines the phenomenon of obesity during pregnancy by tracing the history and policies of maternal weight and nutrition, primarily in the United States. Together these first two chapters provide historical and empirical analysis for understanding why the same solutions to past, present, and future maternal health are reproduced despite significant advances in science and technology over the past half-decade.

I examine this particular constellation of scientific concepts, methods, and health concerns from a feminist and critical race lens. To do this, the analysis required particular methodological attunements like the ones described in the introduction and this chapter, and as summarized in table 1. The methods and concepts that I have developed are aimed at examining seemingly "neutral" spaces like clinical trials. In so doing, my intervention reveals the power relations that remain potent and protected because of systematic obfuscation.

Attending to the politics of power relations illuminates how certain interpretations of "new" science are framed through enduring epistemes and sustain familiar ideologies inherent to race, gender, and individualism; spotlighting these politics can be a path toward reimagining alternative solutions. Imagining alternative solutions to maternal health interventions related to metabolic syndromes requires an intervention in existing epistemes that undergird maternal health interventions, which requires a different set of questions: Who has authority to design trials? Who should be included at the designing table? How do narrowly defined forms of diversity and inclusion efforts actually obscure the impacts of racism on health? What kinds of data should count? How capaciously can we reimagine environmental variables beyond individual pregnant bodies?[106]

*Table 1.*   Key Terms, Methods, and Concepts

| | |
|---|---|
| *Developmental origins of health and disease (DOHaD)* | • This field of research examines how environmental exposures during critical periods (including pregnancy) can influence health across the life course from fetus, to infant, to childhood, to adulthood.<br>• This field of research often draws on epigenetic science. |
| *Epigenetic foreclosure* | • This is a way to capture the unfolding of epigenetics in ambiguous ways without predetermining an outcome. At the edges of foreclosure, there are possibilities for how epigenetic interpretations and applications can be otherwise. |

Table 1.  (continued)

| | |
|---|---|
| *Epigenetics* | • This biological mechanism and field of science explores how gene-environment interactions can impact genetic expression and regulation.<br>• Environmental or behavioral epigenetics are often linked with DOHaD research. |
| *Epistemic environments* | • This concept refers to the ideas and logics that shape contexts of scientific knowledge production, particularly in a twenty-first-century postgenomic era.<br>• These environments include racism, heteronormativity, capitalism (specifically racial-surveillance biocapitalism), late/neoliberalism, and future-oriented or speculative frameworks of value and risk. |
| *Obesity during pregnancy* | • This is defined as having a BMI of 30 or more at the start of pregnancy<br>• This maternal health concern congealed in the 1990s and became a focal point for the integration of DOHaD and epigenetic research in pregnancy trials. |
| *Politics of postgenomic reproduction* | • This feminist framework and tool is conceptualized here to examine the convergence of older biopolitical and reproductive politics that simultaneously endure and are transformed by the emergent postgenomic landscape of science and technology. |
| *Postgenomics* | • This scientific research area shifts away from gene-centric understandings of inheritance and includes the biotechnology of biomarkers and gene-environment research.<br>• Together, epigenetics and DOHaD characterize different aspects of the postgenomic landscape in the twenty-first century.<br>• The *postgenomic era* often refers to the period following the completion of the human genome project at the turn of the century. |
| *Racial improvisation* | • This praxis studies a phenomenon like race/racism that can shapeshift at different rates and becomes relationally reconfigured across micro- and macroscales. |
| *Racial-surveillance biocapitalism* | • This emergent constellation of twenty-first-century capitalisms, as theorized here, considers how together racial surveillance and biocapitalisms create the conditions in which contemporary science and technology are employed to prospect and excavate biological and behavioral data for the production of surplus value.<br>• The convergence of these forms of capitalism uniquely shapes the politics of postgenomic reproduction. |
| *Racist environments* | • This term is developed here as a way to frame exposure to racism as discursive, experiential, material, and visceral. |

## 2 Un/Altered

It is 1950, you're pregnant, and your doctor tells you that if you gain too much weight, you could die.

It is 1970, you're pregnant, and your doctor tells you that you need to gain *at least* twenty-five pounds, or else you'll have a baby that's too small.

It is 1990, you're pregnant, and your doctor tells you that you are overweight and you should only gain fifteen to twenty-five pounds during pregnancy, or else you'll have a baby that's too big.

It is 2010, you're pregnant, and your doctor tells you that you are overweight and you should only gain half a pound per week, *or less*, during the rest of your pregnancy, or you'll have a baby at risk of developing diabetes and obesity.

### INTRODUCTION

Scientific justifications for maternal weight and nutrition recommendations have shifted over time in the United States, but the past half-century reflects a consistent theme: target individual pregnant bodies to manage multidimensional maternal and infant health outcomes. The specific iteration of this theme over the past three decades has been: gaining too much weight is unhealthy, irresponsible, and detrimental to you, the fetus, and your future child. In conjunction with the last chapter, this chapter shows how the science around maternal weight and nutrition has changed and has integrated aspects of epigenetics and developmental origins of health and disease (DOHaD) research, yet the imagined solutions to obesity during pregnancy remain unaltered. I make the case that the durability of lifestyle interventions in addressing obesity and diabetes is deeply connected to the reliance of standards that are based on measuring and evaluating *individual* bodies and behaviors. What results is the persistence of individually-based solutions to multidimensional illnesses.

Here I explore the intellectual and historical processes involved in making weight a measurable object that links pregnant bodies and behaviors to fetal and infant health outcomes, particularly the technoscientific choreographies of standards that shape pregnant bodies. I underscore how key standards like gestational weight gain (GWG) and body mass index (BMI) became the building blocks for maternal health policies and prenatal interventions on obesity during pregnancy. GWG is a standard for assessing the "healthy" or recommended amount of weight gain during pregnancy based on BMI, and it is a unique standard created in the United States. BMI is a vital standard in the creation of current maternal health recommendations, interventions, and treatments of obesity during pregnancy. BMI is a measurement of a person's weight divided by their height; it is a widely used and problematic indicator of health.

In the most recent US guidelines developed by the Institute of Medicine (IOM), the recommendation is that pregnant people should gain a prescribed amount of weight throughout gestation, based on their BMI at the start of pregnancy. This means that pregnant people with a BMI of 25 or higher (classified as overweight and obese) at the start of their pregnancy should only gain half a pound per week, *or less*, during the second and third trimesters. For bodies organized into the "normal" BMI category, the recommendation is to gain about one pound per week in the second and third trimesters. The total recommended GWG for someone who is overweight (or obese class 1) is around twenty-five pounds; for someone who is obese (or obese class 2) it is around fifteen pounds. As of 2013, the data show that around 75 percent of pregnant women gain significantly more gestational weight than the IOM recommends.[1]

I focus on the US IOM guidelines and reports because they provide a time-stamped understanding of obesity during pregnancy in 1990 and 2009, and they are widely adopted recommendations in international settings, except for the United Kingdom. Tracing how and why national recommendations promote standard measurements of weight like BMI and GWG during pregnancy offers a window into understanding the incorporation of epigenetic theories in the etiology of obesity during pregnancy. In the decades between the 1990 and 2009, IOM reports, epigenetics and DOHaD research circulated in the maternal health fields. The 1990 report reflects the growing concern with the "obesity epidemic," and the 2009 report reflects the integration of epigenetics and DOHaD into the most recent conceptualization of obesity during pregnancy. In my analysis of these reports I found that although epigenetic science shifted significant understandings of obesity during pregnancy, the resulting recommendations did not change.

The focus on lifestyle interventions and the use of BMI and GWG as indicators of healthy pregnancies are *still* the same. In 2020, the National Academy of Medicine (NAM) published a discussion paper reviewing the 2009 IOM guidelines that focused on GWG and BMI. The authors find that there is not enough information to justify revisions of the 2009 IOM recommendations, and that there are too many existing gaps in the literature. Consequently, the recommendations from the 2009 IOM report still stand.[2]

Although GWG is not recognized as a legitimate maternal health standard in the United Kingdom, it defines current maternal weight recommendations in the United States and other countries that draw from the United States as a model for their respective health policies.[3] For instance, countries in South America are applying the category of GWG, developed solely in the United States, along with BMI, for the assessment of healthy pregnancies. The United Kingdom has its own recommendations, published by the National Institutes for Health and Care Excellence (NICE). The most recent NICE guidelines from 2010 do not promote or endorse the use or surveillance of GWG as a standard measure for healthy pregnancies. The United Kingdom does not have any GWG recommendations, and the current guidelines do not recommend regularly weighing women during pregnancy.[4]

Emphasizing the role that standards play in knowledge production renders the work that they do visible.[5] Science and technology scholars bring attention to how charts, standards, and recommendations have agency in shaping not only bodies, but also behaviors and approaches to child rearing, eating, cooking, and social relations.[6] The use and development of existing standards are integrated into the clinical translation of epigenetic ideas. This relationship between existing standards and their histories of development in maternal health and nutrition is directly related to the associations made between dietary behaviors in the present and the risk of future disease.

Building on chapter 1, this chapter provides empirical material on how epigenetics and DOHaD reconceptualized the etiology of obesity during pregnancy, and how this reconceptualization did not significantly change the design of current prenatal interventions that focus entirely on individual lifestyle changes to address multidimensional diseases. I argue that BMI and GWG are standards that enable a focus on individual lifestyle changes. The maintenance of these standards across time provides enduring practices, measurements, and frameworks that support lifestyle interventions despite fundamental changes in reproductive sciences. In part, the familiar aspects of nutritional pregnancy interventions, from the 1950s to current prenatal trials, are due to the idea that maternal weight and diet are individually controlled variables that can be measured and examined through standards like BMI and GWG. "Mundane" objects of knowledge production, like charts and standards, are important for understanding larger debates in science and technology studies, such as how does science change, or rather, how does it not change?[7]

Prioritizing the collection, measurement, and assessment of weight alone impedes scientific creativity in finding alternative solutions to obesity. Put another way, the lack of imagination in addressing obesity is due in large part to how static and unimaginative the measurements, data, and standards are in evaluating and framing the problem of obesity. Despite paradigmatic changes in scientific knowledge regarding environmental epigenetics and the known impacts of the social determinants of health, BMI remains a major indicator of health, and lifestyle interventions focused on weight loss (strictly defined as calories in/out) endure as the main mode of intervention. In health-care landscapes across the United States and United Kingdom, there is an unequal dissemination of existing resources. Within such contexts, the public health investment in (ineffective) individual lifestyle interventions is an issue of social justice and health equity.

To address the significance and enduring value of weight in obesity during pregnancy, I explore the historical context of approaches to, measurements of, standards for, and recommendations about maternal weight and nutrition since the 1950s. Then I do a close reading of the charts and content of the 1990 and 2009 IOM recommendations for obesity during pregnancy to show how epigenetics changed the etiology of obesity during pregnancy but did not ultimately change current health recommendations about maternal obesity that overemphasize individual-level interventions.

## HISTORY OF MATERNAL NUTRITION AND WEIGHT

### The 1950s: Maternal Weight Gain and the Risk of Death

> *If you gain too much weight during pregnancy, you could die* . . . (paraphrased from a physician's journal in the 1950s).

To understand the significance of maternal weight in assessing health, I focus on the second half of the twentieth century for two significant reasons. First, with the institutionalization of disciplines, policies, and practices in the postwar 1950s, there emerged a focus on fetal medicine, maternal health, and prenatal care. Second, as I explained in chapter 1, the conceptualization of nutrition as a kind of environmental factor that can influence epigenetic modifications and fetal development in utero cohered at the end of the twentieth century. Together, epigenetic science, DOHaD, and obesity during pregnancy as a health crisis provided the urgent justification for nutritional interventions during pregnancy and heightened surveillance of women's diets and weight before, during, and after pregnancy. Current prenatal lifestyle interventions are similar in content to those in the mid-twentieth century in that they focus on weight and nutrition control.

In the 1950s, the numerical measure of weight gain during each week of gestation, not the nutritional value of the food consumed, was the primary focus of medical interventions in maternal nutrition and health across the United States and the United Kingdom.[8] Doctors and scientists believed that rapid weight gain over one or two weeks of pregnancy was associated with preeclampsia, a condition marked by high blood pressure and protein in the urine that is still linked to preterm birth and high infant and maternal mortality.[9] Reflecting on his medical school training during this period, one doctor stated that if a woman gained two pounds per week, they thought she "was about to die."[10] The scientific consensus at the time was that monitoring and restricting women's weight gain would reduce maternal mortality.[11] This association between preeclampsia and weight gain resulted in the proliferation of medically recommended diets during pregnancy.

Another popular theory that circulated in the 1950s was the "perfect parasite" theory, which framed the fetus as impenetrable and protected from the behaviors, weight, nutrition, or stress of its gestational host.[12] It was imagined that the fetus, as a "perfect parasite," sucked the nutrition that it needed from the pregnant body and received any necessary nutrients through the "maternal instinct." That is, the pregnant person would know what to eat through an instinctual connection to the fetus. This is often associated with the notion of "pregnancy cravings," which assumed that the desire to eat odd foods was dictated by the fetus. For example, if a pregnant person did not eat enough protein, the fetus would compensate by absorbing certain amino acids from maternal blood. Thus, the "perfect parasite" theory assumed that the maternal diet and weight did not matter because the fetus would be able to absorb and direct what it needed from the pregnant body.[13] Drawing from this theory, the prenatal nutrition interventions of the 1950s, aimed at reducing preeclampsia, assumed it was safe to apply calorie restrictive diets during pregnancy.

The imagined maternal-fetal relationship proposed by the "perfect parasite" theory reflects an individualized, unilateral relationship that places the fetus in a position of control to absorb nutrients from its gestational host. This framing of the maternal-fetal relationship was later flipped with the rise of fetal subjectivity and epidemiological risk in the second half of the twentieth century.[14] Medical knowledge about maternal health and nutrition shifted focus to how the pregnant body shaped the risk and the viability of the fetus. As a result, more surveillance is placed on the pregnant body and behaviors as individually responsible for the health and development of the fetus.

### 1970s: Policy Changes in Maternal Health and Nutrition

In the 1970s there was a shift away from the theory of the "perfect parasite," in which the fetus was seen as impervious, toward a recognition of the link between fetal health and maternal health and the endorsement of calorie restriction during pregnancy. Key research findings in the United States and the United Kingdom led to new approaches to maternal nutrition and weight. First, various medical studies verified that GWG did not cause preeclampsia. Second, studies like the Motherwell intervention in Scotland found that pregnant women's diet and weight gain during pregnancy were complexly related to birth weight and brain development.[15] Third, the emphasis on birth weight as a significant indicator of infant health was also motivated by a fear of reproducing "small babies" at risk of infant death.[16] The fear of babies being born small placed the burden on pregnant women to adjust their eating and exercise behaviors in alignment with political and scientific trends that declared specific amounts of GWG.[17] Finally, the increasing rates of maternal and infant mortality leading up to the 1970s resulted in programs to enhance prenatal care across the United States and United Kingdom. Social welfare programs like Medicaid were enhanced to provide prenatal care to uninsured pregnant people. These efforts resulted in the creation of the Women, Infants, and Children (WIC) Program in in the United States. In the United Kingdom, the National Health Service (NHS) further expanded its existing emphasis on maternal health care.[18] These shifts centered the fetus and future child's health as the focus of maternal nutrition, an approach that endures in contemporary prenatal interventions.

The IOM in the United States published a report in 1970 stating that women should gain a minimum of twenty-five pounds of weight during pregnancy.[19] This approach was in direct opposition to the policies and practices promoted in the 1950s that sought to restrict GWG. It was during this period that the notion of "eating for two" became a popular message to pregnant women.[20] The formal recommendation for pregnant women to gain a specific amount of weight in the 1970s helped solidify the focus on, measurement of, monitoring of, and intervention in birth weight and GWG.

From 1970 to 1990, research and national policies in the United States and United Kingdom were aligned on the idea that maternal weight was associated with birth weight, and that low birth weight was associated with infant mortality—though in practice these findings were interpreted differently in each country. US government recommendations and physicians continued to monitor and survey prepregnancy weight and GWG to meet the national guideline of at least twenty-five pounds. Following the UK recommendations, physicians and midwives did not continue to monitor GWG as they had in the prior two decades, nor did they establish an exact amount of weight to gain during pregnancy. British doctors and midwives eventually stopped routinely weighing women and monitoring weight during pregnancy. A veteran midwife whom I interviewed from the UK trial explained that "by the late 1980s they started throwing the scales out of the [clinic] room."[21] Today, the United Kingdom still maintains that GWG is not a dependable standard of healthy pregnancies, but the United States still promotes and applies GWG in current IOM guidelines.

In the 1980s and 1990s, the institutionalization of neoliberalism and social conservatism in the United States and United Kingdom, respectively, shaped the milieu of obesity and maternal nutrition research writ large.[22] Neoliberal policies are characterized by disinvestment in public health care, heightened focus on responsibilization of individuals, and a weakening of social safety nets.[23] Economic recession and a resurgence of social and political conservatism led by Ronald Regan helped instigate disinvestment in Medicaid and other social services. The idea that poor people were getting "fat off the state" was exacerbated by the racist images of "the welfare queen."[24] These racist discourses turned public sentiment away from social programs and impacted reproductive health in the United States. A significant portion of American women covered by Medicaid during pregnancy in 1975 were no longer covered in the 1980s.[25] The growing anti-abortion movement was also influential in the cutting of government spending on reproductive health.[26] Similarly, in the United Kingdom in the late 1980s, Margaret Thatcher's brand of social conservatism resulted in the complete restructuring of social services. In 1990 the NHS and Community Care Act declared the NHS an "internal market," which encouraged competition and converted health providers into NHS trusts, or organizational units with specific functions within the NHS.[27] Some argue that these changes increased regional disparities.[28] The restructuring of the NHS impacted prenatal and postnatal services, resulting in fewer visits with midwives and inconsistent care by different providers. Scholars examine these political shifts as the instantiation of neoliberal policies in healthcare infrastructure.[29]

*1990: Gestational Weight Gain (GWG)*

Whereas the 1970 recommendations were influenced by the fear of "small babies" being born, the 1990 recommendations reflected a fear of "large babies" being born.[30] The 1990 IOM report explained that "gestational weight gain has nearly doubled during the past 50 years—from 15 pounds in the 1930's to a range of 25-35 pounds in the 1980's."[31] The underlying message in the 1990 report was that too much weight gain during pregnancy resulted in babies that were too large for gestation, which is medically defined as macrosomia.[32] The report also associated higher rates of pregnancy complications like emergency cesarean with macrosomia.

Drawing on existing, albeit limited, data, the US IOM report found that since the 1970s white women gained more weight during their pregnancy and had larger babies than in previous decades, and so did Black women, but only slightly. There were no data for any other ethnic or racial group of women. This was not an oversight by the IOM report, but rather a reflection of racialization in the United States at that time.[33] The 1990 IOM report called for more studies on diverse populations, and eventually these calls were enacted into policy; in 1993 the NIH mandated the recruitment of ethnically diverse populations into study designs, a topic I explore further in chapter 3.

The 1990 report connected GWG and BMI standards by claiming that gaining too much weight during pregnancy leads to obesity in the future "mother." This matters for understanding how certain standards continue to remain significant. By connecting GWG to BMI, the use and application of GWG is buttressed by the growing use and application of BMI. Following this causal association between GWG and BMI, the 1990 IOM report recommended that pregnant bodies with a "low" BMI should gain twenty-eight to forty pounds, "normal" BMI was associated with twenty-five to thirty pounds of GWG, and people with a "high" BMI should gain only fifteen to twenty-five pounds during pregnancy.[34]

It is significant to note that these BMI categories were based on Metropolitan Life Insurance's 1959 data. The body mass data collected by this company were not based on any diverse sample size, nor did they include pregnant women. These data were based on the bodies of young, primarily white men from the military.[35] The origins of the BMI standard further reflect how knowledge production was not a "social activity" that included everyone, as some philosophers of science have argued.[36] Rather, the application of exclusive study designs and homogenous datasets was used to create hegemonic standards like BMI, standards that are now widely applied as an indicator of maternal health. The material, methods, data, and designs from the past were packaged into standards that we maintain in the present. Even though the categories of BMI have shifted over the decades to include more data to create specific weight ranges and classes, the inherent structure of the standard—one that focuses on measuring individual bodies and behaviors—remains the same.

Another contribution of the 1990 report is that it defined maternal nutrition and weight based on energy balance, or calories in and out. From this conceptualization, weight loss is a simple equation: more calories out than in. The focus is on the number of calories, not on the quality of the food or how food is metabolized (similar to approaches in the 1950s).[37] Framing GWG in terms of calories in/out makes it something that is *individually controlled*.[38] Despite recent developments in theories of metabolism that suggest more complex framings of weight loss/gain and the role of the environment in shaping fat deposits, US maternal nutrition policies continue to focus on counting and restricting calories.[39] Indeed, the calories in/out approach defines the nutritional intervention in the US SmartStart trial, and the most recent IOM recommendations (2020) for maternal weight and nutrition. Although the United Kingdom did not use GWG, it did depend on BMI, and in both the UK and US trials, interventions focused entirely on individual lifestyle changes to count steps and grams of saturated fats. The production of standards related to GWG, BMI, and birth weight is also crucial to the maintenance of certain scientific and health agendas.[40] In tracing the narrative of maternal nutrition and weight from the 1950s to the 1990s in the United States and the United Kingdom, I argue that these standards have been vital to the endurance of individual lifestyle interventions during pregnancy, a hyper-focus on surveilling pregnant bodies and behaviors, and the pervasive focus on maternal weight in health policy recommendations.

Despite paradigmatic changes in science—such as reconceptualization of inheritance that is not gene-centric, advances in metabolomics, the mapping of the human genome, and postgenomic technologies—it is possible to find similar kinds of prenatal nutritional interventions in the 1970s and in 2015, for two specific reasons: the building blocks, or standards, used to assess the efficacy of prenatal nutrition interventions have not changed, and these standards facilitate the popular use of individual lifestyle changes. GWG and BMI standards are the building blocks of current prenatal interventions. Moreover, the application of GWG and BMI facilitates the production of individual behavioral interventions, not environmental ones, which is politically and economically aligned with late liberal, neoliberal approaches to health care. Counting calories and collecting weight measurements are individually based behaviors that prescribe individual lifestyle changes. Despite the knowledge that epigenetics makes our past, present, and future bodies vulnerable and malleable to unpredictable environmental changes, current prenatal trials focus on surveying and controlling pregnant people's diet and exercise.

In what follows, I focus closely on reviewing the charts and content of the IOM reports from 1990 and 2009. Doing so reveals how measurements and standards focused on weight remain salient in health recommendations, regardless of paradigmatic shifts in scientific knowledge. Such conditions explain the durability of lifestyle interventions in addressing obesity during pregnancy and the ultimate lack of imagination in finding alternative solutions to multidimensional illnesses.

## THE LIMITS TO A PARADIGMATIC SHIFT: 1990 TO 2009, IOM REPORTS ON MATERNAL NUTRITION AND WEIGHT, UNITED STATES

### 1990 IOM Report: Finding Causation

The stated aim of the 1990 IOM report was to identify a "potential causal relationship" between nutritional interventions, GWG, and health outcomes for mother and child.[41] This aim is directly reflected in the diagram in figure 1. This diagram only includes the factors that the 1990 IOM report deemed important, which excludes many environmental and genetic considerations.[42] The diagram indicates that GWG leads to obesity during pregnancy and ultimately to obesity in the mother only and *not* the infant.[43] The schema shows that GWG is also somehow linked to the neurocognitive development of the infant, though the report does not specify how or why this occurs. Data to support these causal associations between GWG and maternal/infant outcomes came from studies completed in the 1960s, including reports from the US Surgeon General's Advisory Committee on Smoking and Health.[44] Using studies that are decades old to inform contemporary health policies and recommendations reflects the well-documented lag that exists between scientific discoveries and their translation and dissemination into practice.[45]

There are three main aspects to highlight in the 1990 diagram: it does not include an environmental category, it represents the maternal-fetal relationship as unidirectional and causal, and it focuses on short-term (and not intergenerational) risks of GWG.[46]

The diagram in figure 1 begins with a specific set of "maternal factors" that can affect GWG and cause obesity. The diagram only addresses "nutritional, sociodemographic, and behavioral" maternal factors, but subsequent sections of the same report explain that there are other general "maternal factors" that can affect GWG and cause obesity during pregnancy.[47] These other factors include "genetic (other than height and ethnic background), health/illness (diabetes, hypertension, chronic disease, systemic or genital tract infection), environmental (geography, climate), and prenatal care."[48] This quote shows that the authors were aware of these other factors, but nonetheless the recommendations focused solely on weight gain as the influential component for obesity etiology. For instance, the role of the environment is essentially absent or limited in the 1990 report. Although the report uses the term *environment* a couple of times, it only references it as "climate and geography." There is no actual description or specific explanation of environmental factors. Thus, the report's recommendations focus on lifestyle interventions to control individual behaviors like weight control, diet, and exercise. The reference to and conceptualization of the environment drastically changes in the 2009 diagram of causality.

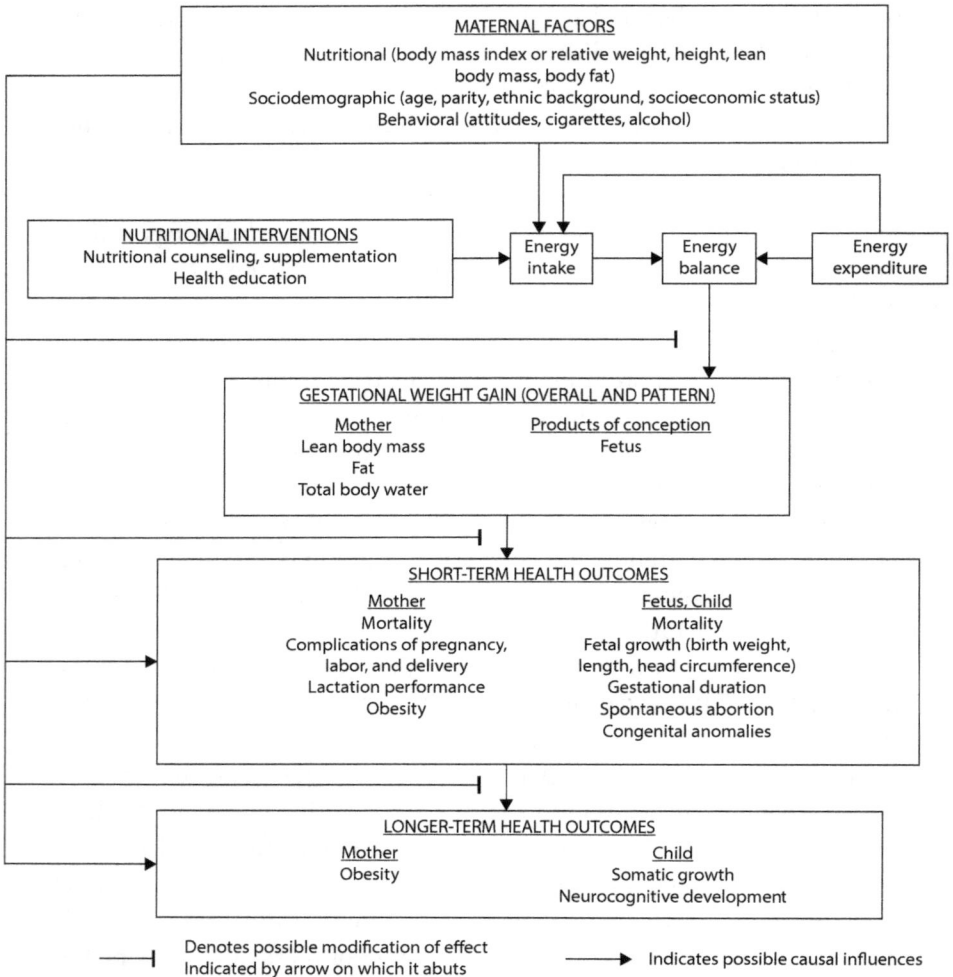

MATERNAL FACTORS
Nutritional (body mass index or relative weight, height, lean body mass, body fat)
Sociodemographic (age, parity, ethnic background, socioeconomic status)
Behavioral (attitudes, cigarettes, alcohol)

NUTRITIONAL INTERVENTIONS
Nutritional counseling, supplementation
Health education

Energy intake

Energy balance

Energy expenditure

GESTATIONAL WEIGHT GAIN (OVERALL AND PATTERN)

Mother
Lean body mass
Fat
Total body water

Products of conception
Fetus

SHORT-TERM HEALTH OUTCOMES

Mother
Mortality
Complications of pregnancy, labor, and delivery
Lactation performance
Obesity

Fetus, Child
Mortality
Fetal growth (birth weight, length, head circumference)
Gestational duration
Spontaneous abortion
Congenital anomalies

LONGER-TERM HEALTH OUTCOMES

Mother
Obesity

Child
Somatic growth
Neurocognitive development

⊢ Denotes possible modification of effect
Indicated by arrow on which it abuts

→ Indicates possible causal influences

*Figure 1.* Schematic summary of potential determinants, consequences, and effect modifiers for gestational weight gain addressed by IOM (1990)

In addition, the 1990 diagram frames the maternal-fetal relationship through unidirectional arrows that flow from the "mother" to the "fetus/child." This emphasizes how maternal factors impact fetal development and does not leave space to imagine the role of multiple other factors that shape fetal development or the role that the fetal genome may also play.

Moreover, it reflects a significant shift away from the "perfect parasite" model that circulated in the mid-twentieth century. Finally, the 1990 diagram frames obesity as a short- and long-term risk for the "mother," but it is *not* a short- or long-term outcome for the "fetus, child." The temporal framing of risk changes dramatically in the 2009 report, when conceptions of intergenerational inheritance and critical periods of development are integrated into the etiology of obesity during pregnancy. Due to the direct influence of epigenetics in the 2009 report, these three areas relating to environment, maternal-fetal relations, and temporal risk alter conceptualizations of obesity during pregnancy. These shifts fundamentally open up and expand the possible causes of obesity beyond maternal weight, BMI, or GWG. Yet these features are selectively ignored in current recommendations.

*Epigenetics and Obesity during Pregnancy: 1990–2009*

Two important contextual shifts in medicine and science occurred between 1990 and 2009. First, the "obesity epidemic" became a global concern. From 1990 to 2000, the number of titles that included the term *obesity* in mainstream media and international health policy reports increased exponentially.[49] In 2000, the World Health Organization (WHO) published a report claiming that obesity was replacing "traditional" global health problems like infectious diseases and undernutrition.[50] The WHO committee also highlighted a glaring contradiction: there are more nutritional plans, exercise programs, and research on weight management than ever before in history, yet the rates of obesity are still climbing in the Global North and South. In addition, obesity research expanded from targeting nonpregnant adults only to children and pregnant adults as well.

The attention and resources invested in the obesity epidemic motivated the national academies of science in the United States and United Kingdom to revise their recommendations on maternal nutrition and weight. In 2009 the IOM in the United States published its report, and in 2010 the NICE in the United Kingdom published its report on maternal nutrition and weight. Both reports reflected the importance of intervening during pregnancy in order to prevent future generations from experiencing obesity and diabetes. The US and UK reports explicitly called for more funding to support the design and implementation of prenatal nutrition trials on overweight and obese pregnant people. The prenatal trials that I study in this book are a direct response to the calls made in the IOM and NICE reports. This is evidenced by the fact that each intervention applies the US and UK recommendations, respectively. The US intervention measures are aimed at controlling GWG, while the United Kingdom does not aim to control GWG, but rather focuses on glycemic control. Yet both trials measure weight and BMI, and each intervention takes the form of lifestyle changes.

The difference in approaching obesity during pregnancy across the United States and United Kingdom materialized most recently in their respective reports on maternal nutrition and health. There is no comparable UK report on maternal weight and nutrition from 1990 like the US IOM report, but the United Kingdom did publish one in 2010. Whereas the 2009 IOM report continued its focus on GWG in relation to BMI, the 2010 report from the UK claims that obesity at the start of pregnancy is more significant than GWG. The UK report states that "if a pregnant woman is obese, this will have a greater influence on her health and the health of her unborn child than the amount of weight she may gain during pregnancy. That is why it is important, when necessary, to help women lose weight *before* they become pregnant."[51] The NICE guidelines support their recommendations by stating that "the US Institute of Medicine guidelines [are] based on observational data" and not evidence-based trials.[52] They go on to say that there are no evidence-based guidelines in the United Kingdom for the focus on GWG.[53] Therefore, there are no recommendations for GWG in the report. GWG does not exist as a reliable measure in the United Kingdom, and in practice UK hospitals do not regularly weigh pregnant people.[54] In contrast, the United States focuses heavily on GWG as a standard that is operationalized in clinical practice.

The other shift that occurred during these two decades is the dissemination and institutionalization of epigenetics and DOHaD studies. As I described in chapter 1, epigenetics and DOHaD draw an explicit connection between women's diets during pregnancy and health risks in future generations. The dissemination of epigenetic science and DOHaD research helped reconceptualize obesity during pregnancy. The 2009 IOM report indexes how notions of the environment, maternal-fetal relationship, and temporality of obesity risk fundamentally changed from the 1990 report.

Despite these new conceptualizations, the 2009 IOM report and rec-ommendations further entrench the reliance on GWG and BMI in current prenatal trials. As I mentioned previously, BMI and GWG are fundamen-tal measures for lifestyle interventions. The dominant use of individual lifestyle interventions to treat obesity and diabetes (across the life span) is indicative of health policies that prioritize individual responsibility and selectively ignore environmental and social impacts on health. Even though new science makes clear that racist and toxic environments unequally impact the rates of diabetes and obesity among poor people of color, this message has not been integrated into contemporary pregnancy trials or maternal health policies. This is epigenetic foreclosure: the pre-vailing investment in individual lifestyle interventions at the expense of structural and systemic change.

## 2009 IOM Report: (Re)producing Obesity

The broad dissemination of ideas related to DOHaD and epigenetics influenced the framing and the data used to create the 2009 IOM report on maternal nutrition and weight, but not the final recommendations. Drawing directly from Barker and others, the 2009 IOM report notes that "developmental programming (or as Barker called it fetal programming) may be mediated through epigenetic events—that is, chemical modi-fications to the DNA and histone proteins that influence gene expres-sion and manifest as phenotypic differences."[55] The report dedicates an entire chapter to developmental programming and the "possible role of epigenetics."[56] In addition, the 2009 IOM committee drew from multiple studies that emphasized a link between environmental exposure, epigene-tic modifications, and the intergenerational inheritance of chronic disease. For the first time in the national history of creating maternal health and nutrition recommendations, the committee included experts in genomics. In 1990, the field of genomics was not referenced or incorporated in the development of the IOM report on maternal nutrition and weight. A word search of the 400-page report did not find the term *epigenetic*, and the term *genetic* was referenced fewer than ten times. In contrast, the 2009 IOM report incorporated literatures and experts from the fields of genom-ics and entire chapters dedicated to epigenetics.[57]

Other aspects of the report are significant. For instance, the committee that was in charge of creating the 1990 report included forty-one people, mostly white, and predominantly based on the East Coast, specifically New York and New England, or on the West Coast, where most of the scholars were from the University of California system and the University of Washington Seattle.[58] The committee formed to revise the 1990 recommendations and create the 2009 report was smaller and included a more ethnically diverse group. Fourteen people were included on the 2009 committee; there was equal distribution across genders; and there was a broader representation of experts from across the East and West Coasts, Midwest, and southern regions of the United States. A handful of participants were included on both the 1990 and 2009 committees, one of whom was also a research collaborator on the US SmartStart trial. While I do not have ethnographic data on how these meetings proceeded, my approach focuses on an ethnographic examination of the content and objects, like graphs, figures, and diagrams that were produced by a nationally organized group of scholars.

The 2009 IOM report justifies the need to reevaluate maternal nutrition and weight in the following excerpt: "It has become clear that heavier women could gain less weight and still deliver an infant of good size. Since [the 1990 report] the obesity epidemic has not spared women of reproductive age. In our population today, more women of reproductive age are severely obese (8 percent) than are underweight (3 percent), and their short- and long-term health has become a concern in addition to the size of the infant at birth. Clearly the time had come to reexamine the guidelines for weight gain during pregnancy."[59]

This excerpt frames obesity in direct relationship to reproduction and draws on the most extreme cases of obesity on the one end and underweight classifications on the other as a justification for the need to reevaluate the guidelines for GWG. It is important to note here that the majority of the population falls in the middle and is not included in the most extreme classifications of BMI. Also, the 2009 report draws from updated BMI categories established by the WHO in 2000. The new categories contain four groups instead of the three groups developed by the Metropolitan Life Insurance group. Figure 2 displays the 2009 IOM causal diagram of GWG and obesity, which is based on the 1990 IOM report. One aspect

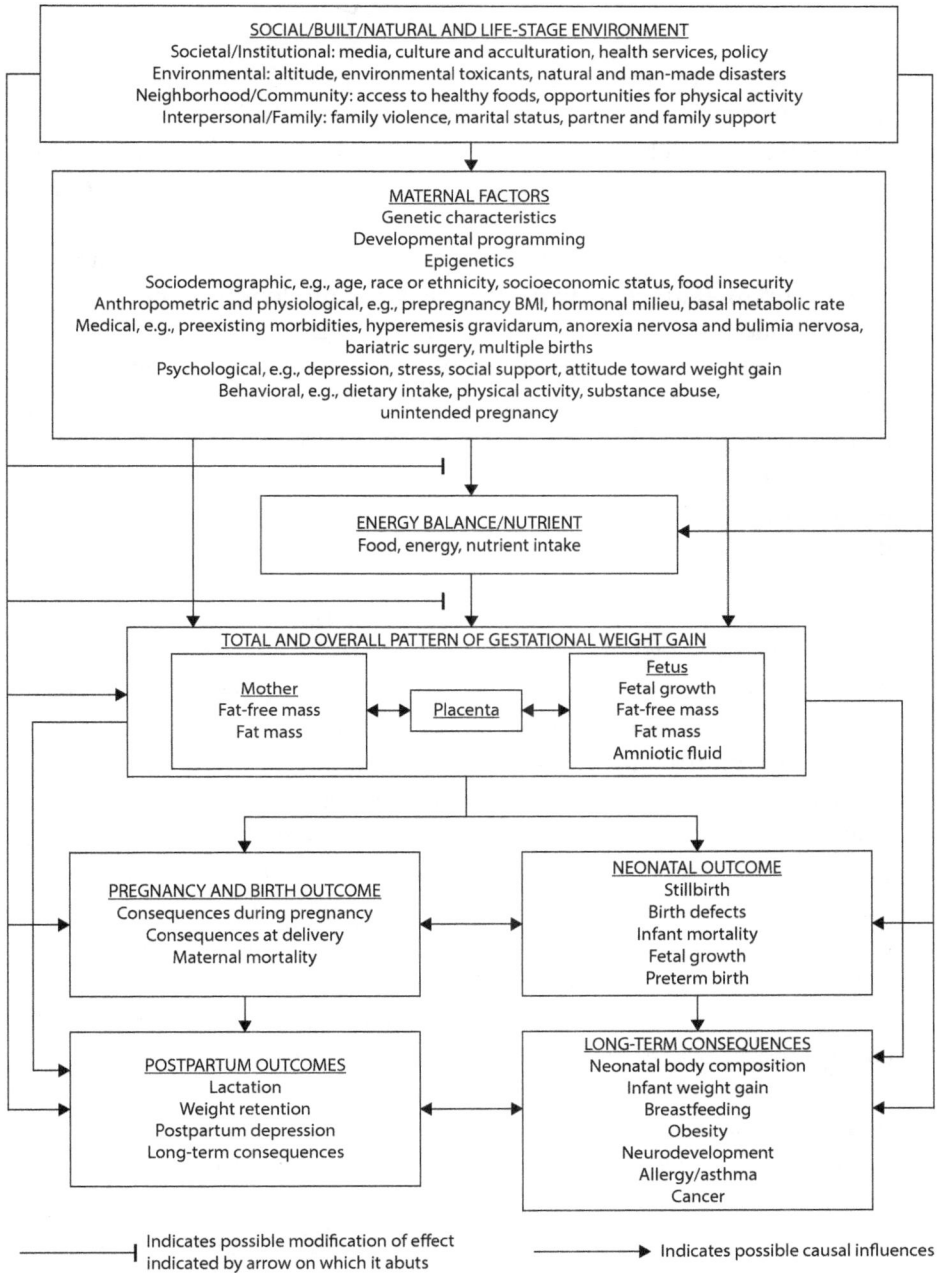

SOCIAL/BUILT/NATURAL AND LIFE-STAGE ENVIRONMENT
Societal/Institutional: media, culture and acculturation, health services, policy
Environmental: altitude, environmental toxicants, natural and man-made disasters
Neighborhood/Community: access to healthy foods, opportunities for physical activity
Interpersonal/Family: family violence, marital status, partner and family support

MATERNAL FACTORS
Genetic characteristics
Developmental programming
Epigenetics
Sociodemographic, e.g., age, race or ethnicity, socioeconomic status, food insecurity
Anthropometric and physiological, e.g., prepregnancy BMI, hormonal milieu, basal metabolic rate
Medical, e.g., preexisting morbidities, hyperemesis gravidarum, anorexia nervosa and bulimia nervosa,
bariatric surgery, multiple births
Psychological, e.g., depression, stress, social support, attitude toward weight gain
Behavioral, e.g., dietary intake, physical activity, substance abuse,
unintended pregnancy

ENERGY BALANCE/NUTRIENT
Food, energy, nutrient intake

TOTAL AND OVERALL PATTERN OF GESTATIONAL WEIGHT GAIN

Mother
Fat-free mass
Fat mass

Placenta

Fetus
Fetal growth
Fat-free mass
Fat mass
Amniotic fluid

PREGNANCY AND BIRTH OUTCOME
Consequences during pregnancy
Consequences at delivery
Maternal mortality

NEONATAL OUTCOME
Stillbirth
Birth defects
Infant mortality
Fetal growth
Preterm birth

POSTPARTUM OUTCOMES
Lactation
Weight retention
Postpartum depression
Long-term consequences

LONG-TERM CONSEQUENCES
Neonatal body composition
Infant weight gain
Breastfeeding
Obesity
Neurodevelopment
Allergy/asthma
Cancer

⊣ Indicates possible modification of effect
indicated by arrow on which it abuts

→ Indicates possible causal influences

*Figure 2.* Schematic summary of potential determinants, consequences, and effect modifiers
for gestational weight gain, modified from IOM (1990) guidelines in IOM and NRC (2009)

that remains consistent across both the 1990 and 2009 diagrams is the focus on how different factors shape maternal weight gain, or GWG.

A significant difference between the 1990 and 2009 diagrams of causation is that the environment exists as a main factor in the 2009 diagram and overall report. The "environment" is defined in a capacious manner to include aspects across "multiple levels" and domains of experiences. In the 2009 report, the environment includes "culture, acculturation, media, toxins, policy, access to health food, [and] family violence" among many other factors.[60] In addition, the report claims that "some of the most significant determinants of GWG at multiple levels (social/institutional, environmental, neighborhood/community, interpersonal/family, and individual levels) occur *across the life course* [emphasis added]." The inclusion of temporality and scale coheres to produce a more complex understanding of how the environment can affect pregnancy. Whereas the 1990 diagram connects the environment directly to maternal factors only, the 2009 diagram links the environment *directly* to multiple categories: "energy balance/nutrients," "total and overall pattern of gestational weight gain," and short- and long-term health outcomes for both mother and child. The expanded and more direct role of the environment in the 2009 report is indicative of postgenomic, epigenetic influence.

Another significant change is in the portrayal of the fetal-maternal relationship. In the 2009 diagram the "mother" and "fetus" are distinct categories, connected by double-ended arrows. Through epigenetics, we can examine the fetal-maternal relationship as one that is dialectical and not unidirectional, one that is vulnerable to environmental factors on different scales. The diagram and the 2009 text itself represent an open "dialogue" between the mother and fetus. The report includes an entire chapter related to the "dialogue between the fetus and the mother."[61]

Furthermore, the maternal-fetal interaction is mediated by the placenta. *The placenta was not included, and essentially did not exist as a significant factor in obesity during pregnancy, in the 1990 diagram or report.* In 2009, the placenta is at the center between the "mother" and the "fetus," and all three categories are connected by a double-ended arrow. The double-ended arrows signify that the direction of "causal influence" can go both ways. In the 1990 schematic there are no double arrows. This conceptual difference in the potential for causal influence to be reversed or to move in more than one direction is also a fundamental change in the notion of causation.

Along with the more open-ended description of the environmental scale and the dialectical relationship between mother and fetus, there is also a temporal shift in risk. In the 2009 diagram, the fetus/developing child is framed as at risk for obesity in the long-term future. The 1990 report associates short- and long-term risk of obesity to the mother only and not the "fetus/child." In addition, the "long-term" risks that were assigned to the "fetus/child" extend only to the "child." In 2009 the temporality of risk is expanded into adulthood, which reflects the focus on disease across the "life course" or "life stage," which is taken directly from DOHaD research. The 2009 long-term consequences for the developing child include obesity, cancer, asthma, and allergies. All of these are *still* associated with the GWG of the mother and broader environmental exposures.

My analysis shows that rather than creating recommendations that consider capacious framings of the maternal environment or comprehensive applications of epigenetics, the solution remains the same: target maternal weight through individual lifestyle interventions. Epigenetic science changed conceptions of permeability between bodies, placentas, and the environment. Nevertheless, the future risk of obesity for children is directly associated with the individual diets and behaviors of pregnant women in the present. This temporal framing of risk selectively focuses on the future risk without acknowledging the indeterminate potential of epigenetic changes. By highlighting the temporal, relational, and environmental differences between the 1990 and 2009 diagrams, I am making visible the impact that epigenetic ideas have had on the reconceptualization of obesity during pregnancy. And yet the 2009 diagram shows how durable the standards of GWG and BMI are for examining and assessing obesity during pregnancy. The postgenomic paradigm altered the etiology of obesity during pregnancy, but how obesity is measured and evaluated still depends on existing standards that have been developed over decades in maternal nutrition and health.

To be sure, the experts on the 2009 committee closely evaluated the existing data, and although there were and are many inconsistencies linking GWG to pregnancy outcomes, the current recommendations claim that pregnant people with a certain BMI should still gain a certain amount of weight during pregnancy.[62] Whereas the United Kingdom did not and does not promote the use of GWG, the United States still remains squarely focused on this standard. Regardless of the distinct recommendations, both national contexts focus on weight through the use of BMI, which is still directly related to individual lifestyle interventions.

## TRACING BANAL KNOWLEDGE OBJECTS

Tracing the scientific and historical context of maternal health policies that focus primarily on maternal weight reveals a variety of shifts in national recommendations regarding obesity during pregnancy in the United States and United Kingdom. In the 1950s, ideas about the "perfect parasite" helped justify restrictive prenatal interventions aimed at minimizing weight gain. In the 1970s, the link between maternal weight and infant outcomes stimulated a different set of approaches: women were expected to "eat for two" and gain at least twenty-five pounds. In addition, the maternal-fetal relationship shifted from one that was unaffected by maternal nutrition and behaviors, reflected through the perfect parasite theory, to one in which individual pregnant women are intervened upon and held medically and legally accountable for fetal and infant outcomes. Then in the 1990s, the United States formally integrated GWG recommendations in relation to BMI into maternal health guidelines.

The historical material of maternal nutrition and health also reflects the divergence across US and UK recommendations. Throughout the second half of the twentieth century, the United States and United Kingdom worked closely together, drawing on similar research and scientific findings to create respective national policies on maternal weight and nutrition. Early on there was mutual agreement on basic findings, yet by the 1990s different interpretations produced distinct clinical practices and eventually distinct national recommendations by 2010.

I mentioned at the beginning of this chapter that in January 2020, the National Academy of Medicine (NAM) in the United States published a discussion paper reviewing the 2009 IOM recommendations. The NAM also concluded that the existing literature on GWG is inconclusive and requires *more research*. Instead of promoting entirely different kinds of prenatal interventions that consider environmental and structural factors, the NAM claims that more prenatal interventions using GWG and BMI are needed.[63] It further specifies that these future studies need to focus on people with higher BMIs and to explore how race and economic status impacts weight— *but not necessarily how intervening in issues of racism and poverty could impact weight.*

The future vision of maternal health and nutrition remains squarely fixed on weight, as an individually controllable variable and unit of measurement. Centering weight means that future studies may continue to focus on lifestyle interventions targeting diet and exercise. Moreover, the past calls for research from the National Academy of Sciences and Medicine in 1970, 1990, and 2009 shaped subsequent funding streams. The implications of the 2020 recommendations is that scarce resources will be invested in the same types of individually based interventions. Durable standards, measurements, and variables of evaluation impact how future study designs can be imagined and which studies are funded.

Juxtaposing the 1990 and 2009 IOM diagrams of causation reveals the changes that emerged as a result of new scientific knowledge: we see clearly that significant aspects of obesity etiology changed, and yet the interventions and recommendations remained unaltered. The conceptual changes that came about through the integration of epigenetic science and DOHaD research, such as new definitions of the environment, maternal-fetal relations, and the temporality of risk in the 2009 report, were not integrated into the recommendations. The analysis of the reports shows that science can effect only certain types of change, indexing an aspect of foreclosure that occurs in the translation of scientific ideas into policy recommendations. Furthermore, theorizing standards like GWG and BMI alongside shifts in new science spotlights how limits to innovation can be found in seemingly low-tech spaces like obesity during pregnancy. Ethnographically examining seemingly banal and commonplace areas of postgenomic reproduction sheds light on relevant debates within science and technology studies, like the unequal and unreciprocal relationship of influence across science and society.

The overdependence on BMI and GWG enables the perpetuation of individual lifestyle interventions, and the perpetuation of individual lifestyle interventions indexes existing social and political contexts. Put another way, the standards and measurements applied in pregnancy trials and promoted in national health guidelines are shaped by the epistemic environments that create extreme social and economic inequality. Despite scientific knowledge that individuals are permeable and vulnerable to environmental factors *outside their control,* and despite the reality that racism and social inequalities create conditions that make it *impossible* to execute or sustain lifestyle interventions, national health organizations continue to call for individually based solutions to multidimensional illnesses. If the ways we measure and frame the problem of chronic illnesses like obesity and diabetes do not change along with the increasingly complex understandings of causality and biological malleability that new science is spotlighting, the promise of epigenetics will continue to be foreclosed. Such instances of foreclosure risk reproducing the same ineffective solutions to future health issues.

# Part II

# 3  Politics of Recruitment

HOW FATNESS, RACE, AND RISK SHAPE
CONTEMPORARY PREGNANCY TRIALS

You are sitting in the waiting room of a prenatal clinic. This is your first appointment to confirm your pregnancy.

This is a private clinic, with art on the walls. The wall sculpture looks like a decapitated woman in the fetal position (see figure 3). You notice how pronounced the rib cage is. You wonder if the art was commissioned as an empowering representation of the "ideal woman": thin, headless, with only the "essential" parts for reproduction. On later visits, you will never really notice this art piece again because it seems to blend in with the whole space. But it is always hanging over you.

Only one other person is in the waiting room with you. She has some pamphlets and brochures in her hand. She approaches you, and you try not to make eye contact, but she starts talking to you anyway.

She says, "Are you interested in improving your pregnancy? If so, call this number and you can get more information about this new trial that helps pregnant women with their diet and exercise." She goes on to say that there's a weight requirement and you look at her, purse your lips, and exclaim, "Are you asking me to participate because you think I'm fat!"

The researcher from the trial looks startled. You get up and walk over to the registration station, and you tell the nurse that you are upset and that this encounter was inappropriate.

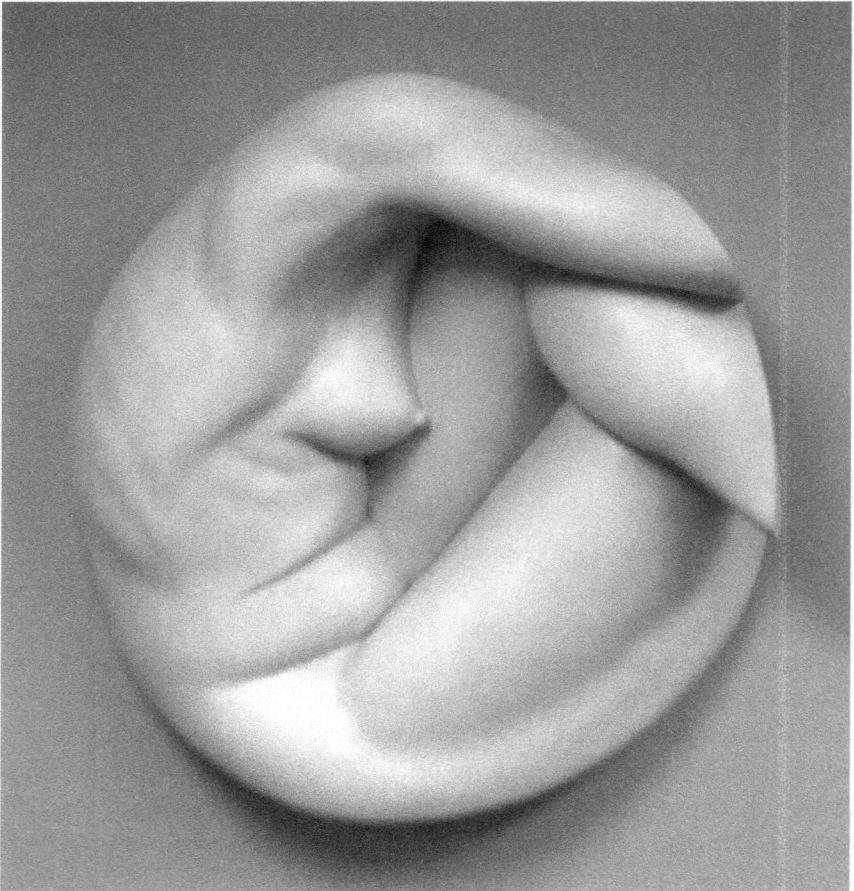

*Figure 3.* Photograph of sculpture from a recruitment clinic on the West Coast, United States. Photo by author.

## INTRODUCTION

This vignette is drawn from an actual recruitment encounter that occurred in the US prenatal trial. Trial eligibility required participants to have a body mass index (BMI) of 25 or higher in the United States, and 30 or higher in the United Kingdom, at the start of pregnancy. Some participants enrolled in the trial because they were concerned about their weight; however, the BMI requirement itself was a barrier for recruitment, as some women declined to participate due to the stigma associated with being overweight or obese.

Recruitment is the most important and laborious phase of clinical trial implementation. Without a precise group of appropriate participants, no clinical trials are possible. The trials I examined required ethnically diverse, overweight or obese, pregnant bodies. This population group needed to be healthy enough—with no comorbidities like diabetes or heart disease—but they also had to meet the overweight and obese BMI requirement, which is often stigmatized as "unhealthy." Most of the existing information on who participates and why in clinical trials exists for pharmaceutical drug trials, not pregnancy trials. There is a dearth of data on the recruitment efforts that target pregnant populations in general, and particularly overweight and ethnically diverse populations. The lifestyle pregnancy trials I examine are classified as behavioral trials, because their main intervention is diet and exercise, not pharmaceutical drugs; consequently, they are deemed safer for pregnant populations.

Recruiting participants for lifestyle pregnancy trials required an immense amount of resources, time, energy, and coordination across multiple levels of each trial. It also presented a variety of challenges across different national settings. Even though nationalized health-care contexts (like the National Health Service in the United Kingdom) facilitated access to larger numbers of participants, challenges to recruitment still emerged. For instance, pregnancy trials in the same urban areas competed for a limited number of pregnant participants. Across both the United States and United Kingdom, the main challenge was to recruit underrepresented minority communities who also met the BMI requirement. For instance, in the larger group of US trials it was difficult to recruit African American and Native American participants, and in the United Kingdom it was challenging to recruit Southeast Asian participants, for reasons that I explore later in this chapter.

Recruitment is the phase in which people are transitioned into research subjects and data points for science. The recruitment of pregnant bodies provides the data necessary for speculating on future health, which I explore in chapter 6. Even though most lifestyle pregnancy trials are inconclusive, they remain valuable for future health because they are sites for the collection and prospecting of pregnant biobehavioral data. The data amassed from these trials support a whole network of scientific careers and private collaborations.

Recruiting any pregnant person into a clinical trial, let alone overweight and ethnically diverse people, is unique because they are not treated, socially or medically, as equal to nonpregnant persons. I emphasize that the discourses of risk used to justify and encourage the participation of pregnant overweight people in lifestyle intervention trials are quite distinct from traditional discourses of risk and consent in trials that target nonpregnant populations. Unlike nonpregnant (primarily healthy) adults who consent to bodily risk in phase I drug trials, pregnant people who are deemed obese are framed as a health risk *before* they enter the trial, and they are framed as a health risk not just for themselves, individually, but also for the next generation. Epigenetic and developmental origins of health and disease (DOHaD) theories help justify prenatal interventions as prevention for future health risks, not exposure to risk in the present. In addition, the epistemic environments of neoliberalism and capitalism have made clinical trial participation an increasingly integral part of health-care services, and these milieus influence dominant conceptions of healthy and risky bodies. Within this backdrop, notions of health risk are also intimately imbricated with fatness and race.

In what follows I examine how the role of fatness shaped recruitment across both national settings, then explore the distinct politics of recruitment in the United States and United Kingdom, respectively. This organization frames the US and UK trials as separate ethnographic case studies.[1] By reading the ethnographic material side by side, I highlight the limits and possibilities of standardizing scientific methodologies in distinct national settings. For instance, this approach helps reveal how each trial was similarly designed, was implemented in a standard way, and experienced similar recruitment challenges. Yet the ethnographic data show how local contexts strongly shape the recruitment of diverse populations. By employing an intersectional lens, I show how notions of fatness, race, and risk shape the politics of recruitment in contemporary pregnancy trials.

## FATNESS AS A BARRIER AND MOTIVATOR TO TRIAL PARTICIPATION

Across the United States and United Kingdom, internal and external fat stigma motivated and discouraged participants to enroll in the trial.

Many participants across the United States and United Kingdom said that they joined the study because they were worried about gaining too much weight during pregnancy. These participants were also told by their doctors that having a high BMI was a pregnancy risk, and that joining studies like the SmartStart or the StandUp trials could help them have a healthy pregnancy. Some participants, as I explore in more detail in the next chapter, wanted a program that would strictly monitor their diets and exercise.

Even when participants had experienced negative medical encounters about their weight in the past, some still found the pregnancy trials on diet and nutrition helpful. One participant in the intervention arm of the UK trial explained to me that she had always been overweight and that "the doctors they just yell at you and have a go at you—I find it patronizing because they assume you know better but it's not straight forward. I didn't know how to eat before, at least with [the trial] they explain how to do it." Her comment illustrates how participants who are strongly interested in receiving extra surveillance of their weight and diet will self-select into the trial.

However, fat stigma in both the US and UK contexts also discouraged participants from enrolling because they felt ashamed or were upset at being targeted because of their weight. Pregnant women, similar to the recruitment encounter at the beginning of this chapter, would refuse to participate in a trial if they thought they were targeted because of their weight. Overweight pregnant adults report that they are initially recognized as fat, and not pregnant, which stirs up emotions of shame and impacts their medical interactions.[2]

Like the US site, the UK site also had recruitment issues due to the BMI requirement. Out of one hundred women approached to participate in the UK trial, only thirty enrolled.[3] In an interview with the social scientist hired to work on the UK trial, she stated that fat stigma was a barrier to trial participation. Based on her interviews with women who declined to participate in the UK trial, she found that overweight women were resistant to engaging with more medical attention during pregnancy. Seventy percent of women declined to participant in the trial because they did not want to be told how to eat or how to manage their weight.[4] The challenges to recruiting overweight pregnant people into lifestyle trials are intimately entangled with the social context of fatness in the United States and United Kingdom.

## Intersections of Fatness

Being fat in the United States or United Kingdom is stereotyped as a sign of laziness, lack of discipline, lack of education, and low economic status.[5] Fat bodies are deemed failures in a neoliberal society that emphasizes individual responsibility and the maximization of labor and productivity.[6] Good citizens, as the rationalization goes, should take care of themselves.[7] Responsible citizens should eat "healthy" and stay fit no matter the socioeconomic policies that make it impossible to buy fresh produce in their neighborhood, safely walk outside, or make healthy dinners each night because they are juggling three jobs. Fat bodies are often pathologized as unhealthy, risky, and abnormal (despite the fact that a majority of the population is classified as overweight in the United States and United Kingdom).[8] Overweight nonpregnant adults report hesitation or anxiety regarding medical encounters for fear of fat shaming.[9] Fat bias in a medical context assumes that someone who looks fat is unhealthy, and that any medical issues they have are due to their fatness, whereas someone who looks thin is not automatically assumed to be unhealthy.[10] For instance, gestational diabetes mellitus (GDM) impacts people across the BMI spectrum, but fat bias affects how certain bodies are tested, targeted, and surveyed.[11]

To complicate this matter further, not all scientists agree that weight or fat alone is the main variable to measure or target for intervention. At an obesity conference that I attended in 2013 with the US principal investigator (PI), there was much debate regarding the use of BMI as a key indicator of health. Some scientists explained that a more accurate measurement to use is body composition, because not all bodies carry fat in the same way, and not all fat is the same. For instance, people who are shorter can jump from normal weight to overweight, or overweight to obese, with a weight fluctuation of two to five pounds. Body builders who carry more muscle weight are technically classified as obese. Regardless of the social and material complexities around fatness and obesity, current studies on prenatal nutrition reduce the issue of maternal nutrition and weight to calories in and out and individual choices.[12]

Public health scholarship frames low-income communities of color as an "at risk" group for obesity.[13] The SmartStart and StandUp trials both draw on the existing public health literature to justify their interventions and target population. Thus, the need and justification to recruit overweight, ethnically diverse participants is symptomatic and indexical of existing enduring relations across race and fatness. Although most public health discourses blame individuals for their metabolic syndromes, people of color do not have poorer health outcomes because of their racial/ethnic classifications or behaviors. Instead, scholars emphasize that the disparities in obesity and diabetes are indicative of long-term exposure to systemic racism, deregulation of food companies, governmental corn subsidies, and national free trade agreements.[14]

Race and fatness come together through an interconnected set of hierarchies. Sabrina Strings's book *Fearing the Black Body: The Racial Origins of Fat Phobia*, throws into relief how the convergence of slavery, religion, and racism in the United States link Blackness with fatness. By tracing the different social and medical approaches to fatness across the nineteenth and twentieth centuries, Strings finds that aesthetics of fatness became gendered, moralized, and racialized.[15] She states that fatness and thinness "have not, principally or historically, been about health. Instead, they have been one way the body has been used to craft and legitimate race, sex, and class hierarchies."[16]

In the early 1900s, US perspectives on public health, founded on explicit eugenic logics, assumed that "degenerate" communities would not survive, and thus investing resources in improving the health of such communities was not necessary. Influential experts in science and medicine (including John Harvey Kellogg, the American symbol of healthy eating, and health reform) believed that "the nonwhite races would eventually die off due to their weak constitutions."[17] While the emergence of diversity and inclusion efforts in public health shifted more research, resources, and attention onto low-income communities of color in the second half of the twentieth century, Strings's analysis reminds us that the link between fatness and Blackness was made prior to the public and moral crisis of obesity. The key point here is to be mindful of how underlying epistemes of eugenics and racism shape past and current health interventions in obesity and diabetes.

Being *pregnant* and *fat* and *a person of color* entails even higher stakes and predetermined risks. Fatness, along with aspects of race, class,

immigration status, and education, all shape who is blamed and targeted
for heightened surveillance and intervention, especially for bodies framed
as reproductively capable.[18] Women's bodies are publicly made available
for social commentary and are framed as part of national and imperialist
projects that require state intervention.[19] For instance, Khiara Bridges's
work documents how Black and Brown pregnant women who receive
prenatal care through Medicaid must follow nutritional plans, and if
they do not comply they can be penalized by the state for fetal neglect
and harm.[20] More acutely, pregnant women with a high BMI are framed
as having "risky wombs or intrauterine environments."[21] The historical,
political, and medical contexts that connect anti-Blackness, fatness, and
(un)healthiness influence people's desires and motivations to enroll in
lifestyle prenatal trials and also influence the justification for the trials in
the first place.


## UNITED STATES, SMARTSTART TRIAL

### Recruitment Infrastructure

After the encounter illustrated at the beginning of this chapter, the recruitment strategy at the US SmartStart trial shifted away from explicitly stating the weight requirement during the initial contact to focusing on how
the trial wanted to help improve pregnancies by making changes to diet
and exercise. Interested participants were contacted over the phone, at
which time the staff clarified the weight requirement. Eligibility was confirmed at the first intake appointment by collecting weight and height
measurements.

I was trained and worked as a staff member in the US SmartStart trial,
recruiting participants and delivering the intervention to participants in
the experimental or intervention group. I gained intimate knowledge of
trial protocols and procedures, and I worked closely with all staff, PIs,
collaborators, and participants in the intervention group of the trial.
I expand more on my experience as an interventionist in chapter 5. My
typical workweek included staff and intervention meetings, consortium
calls, recruitment visits, data entry, and intervention delivery visits. I was
assigned certain private and public clinics within a ten- to twenty-mile

radius from the trial's centralized office. At the weekly staff meetings, which included the PI, project coordinator, research assistants in charge of data collection, and interventionists who delivered the intervention, we spent hours strategizing about recruitment efforts. Three days a week I visited the participating clinics to check in and build rapport with the nursing staff. We even had prizes for the clinics that referred the most people. At times I felt like a salesperson, going from clinic to clinic passing out the trial brochures, waiting to catch the pregnant women after their prenatal appointments. On a good day, I would get a list of eligible names from the nurses; other days I had to do a "cold" approach to anyone who was leaving their appointment. Some women would come out of their appointment feeling extremely overwhelmed and would rush out of the clinic. In these situations, standing next to the door was often the only point of interception. In the rushed scenarios I tried to hand off the brochure to them, then hoped they would look at it and follow up with us on their own.

In other instances, I would strike up a conversation with women in the waiting room and introduce them to the trial in a calm manner. In this scenario, the "recruitment" encounter could take longer because women would share all sorts of stories with me. One woman explained her first few miscarriage experiences and the challenges she faced with her family and doctors. While her doctors treated her miscarriage as a routine reproductive process (one in five women experience miscarriages), her family blamed her for not being fit enough to carry to term, as if miscarriages were voluntary.[22] At other times, the conversation would stray away from the topic of pregnancy altogether. One woman talked passionately about her activist work with incarcerated communities and her philosophy on capital punishment (she almost recruited me to volunteer at a nearby prison).[23]

The US prenatal trial that I worked on was part of a multisited national consortium funded by various branches of the National Institutes of Health (NIH), including the National Institute of Diabetes and Digestive and Kidney Diseases (NIDDK), National Heart, Lung, and Blood Institute (NHLBI), and National Institute of Child Health and Human Development (NICHD). The SmartStart trial was awarded over $3 million in funding between 2012 and 2017. Recruitment took place during

2012–2015, and I worked on the trial in 2013 during a time of intense recruitment efforts.

The consortium included seven individual trial sites that also worked collectively to collect similar data. For instance, each site collected "common core" measurements or data (such as weight, calories, steps, and biosamples), which were standardized across the sites and were then pooled to increase the overall sample size of pregnant overweight people. All the sites implemented a randomized controlled trial (RCT) of a lifestyle intervention with overweight, pregnant women. All sites focused on reducing gestational weight gain (GWG), but different sites applied different intervention components like online programs or meal replacement shakes.

The consortium included 1,150 participants in total across all individual trial sites. The SmartStart trial recruited around 250 participants. Any trial sites that did not meet the monthly recruitment numbers had their funding cut off. Two of the trials ended prematurely because they did not recruit enough people. To ensure that the SmartStart trial met the sample size goal, there was a partner site on the East Coast. Unlike the West Coast site, the East Coast site was located in a research institution with an affiliated hospital.[24]

### Why Do People Participate in the US Prenatal Trial?

Participants were motivated to enroll in the US trial for many different reasons. The motivations to participate in lifestyle pregnancy trials are somewhat similar to those for the broader nonpregnant North American adult population, who enroll in clinical research primarily for the monetary incentive and to improve science.[25] The participants in the US SmartStart trial were paid $25 for five assessment visits (or data collection visits), during pregnancy and postpartum. In the last visit, at the one-year postpartum follow-up, they received $50. In total, the participants could potentially earn $175 for their entire participation.[26] The trial protocol framed the monetary incentive as a way to "promote recruitment" and to ensure "retention." This amount of money is far less than the financial incentive in phase I drug trials, which can pay participants thousands of dollars.[27] In contrast to the broader research on nonpregnant adults who participate in clinical research, pregnant participants in the United States did not mention or refer to altruism or the desire to improve science as a key motivating factor to enroll in the trial.

From my observations and my own work as a staff member recruiting for the trial, the monetary incentive was significant for the women who were unemployed at the time, but it was not the only reason women participated. In addition to the small financial incentive participants receive to enroll in the US trial, the underlying contexts of risk associated with pregnancy and fatness also shape individual motivations to participate. In an interview with the PI, I asked her why she thought pregnant women volunteered to participate in the trial. She responded: "I think it is women who care about their health and their babies' health. [. . .] I think some join the study because they think they should, because they heard about it in their doctor's office. [. . .] Some do it because they are really worried about gaining weight, some do it because they are worried about their baby, some do it for money. I think they have all different reasons for joining, but my hope is that they are doing it for their health, so that is my motivation but I don't think that it is necessarily theirs."

Thirty-eight percent of the people in the targeted recruitment area earned less than $25,000.[28] Since I spent more of my time recruiting and delivering the intervention at the public rather than the private clinics, I observed many participants from a lower economic class. The trial protocol defined the patient population at the public clinics as more ethnically diverse, of lower socioeconomic status (SES), and younger compared to the private clinic population, which had older patients with high SES. Although the trial aimed to recruit an equal number of participants in different income ranges, this proved difficult in the implementation of the trial. In general, it was hard to recruit participants due to fat stigma, and it was also harder to recruit from private clinics for a variety of reasons related to insurance systems, access, and attitudes toward clinical experimentation.

*Recruiting Minorities in Clinical Trials*

A key reason for collaboration across the national consortium also had to do with the diversity of populations. The Revitalization Act of 1993 requires clinical trials that are funded by the NIH to include racial/ethnic minorities and women.[29] The US consortium that housed the SmartStart trial included sites in California, Puerto Rico, New York, Missouri, Illinois, Arizona, and Louisiana. These sites were each responsible for recruiting what they termed "Hispanic, African American, Native American, and Caucasian" populations. In the consortium meeting calls regarding recruitment, it became clear that the most challenging groups to recruit were African American and Native American populations.[30] In these conversations, no one mentioned the history of unethical trials with these populations, namely the Tuskegee syphilis experiment conducted on African American men for over thirty years (despite the existence of a known cure), and the unethical experimentation on Navajo uranium miners in the 1950s. In addition, there was no mention or acknowledgment in any meetings or protocol descriptions related to recruitment regarding the unethical testing of birth control drugs on mentally ill patients and on poor women in Puerto Rico.[31] The historical context of medical and experimental exploitation disappeared from view in the recruitment phase of contemporary pregnancy trials.

At the SmartStart trial on the West Coast site, the majority of the women I worked with were women of color. The demographics of the West Coast site included large Mexican, Guatemalan, and Indigenous communities. Forty percent of the population in the main recruitment area of the trial was "Hispanic."[32] The target enrollment for the trial was 50 percent "Hispanic" and 50 percent "non-Hispanic women," and these classification terms were used in the study protocol without justification. In conversations with staff members the label "Mexican American" was often used interchangeably with "Hispanic." The trial also recruited some Afro-Latinas, which further complicated recruitment classifications.

*Hispanic* and *Mexican American* are not comprehensive terms. There is great diversity *within* the homogenous category Hispanic, which is rarely studied in public health research. There are also racist tensions between and among Mexican American populations across class, status, and indigeneity.[33] These nuanced notions of diversity are not considered in the application of ethnicity classifications in the trial. However, they came up in the process of implementation, particularly around language translation; the trial was only equipped to translate the study into Spanish, and not all Latin American Indigenous immigrants speak Spanish.

For instance, in the state of Oaxaca, Mexico, there are more than sixteen Indigenous languages, and there are large Oaxacan immigrant communities on the West Coast that speak Mixtec and Zapotec.

In their work on race in human genetics research, Hunt and Megyesi confirm that the use of ethnic and racial categories is often left unquestioned and not rigorously researched in the design of the projects, resulting in different scientists applying categories of race and ethnicity in wildly different ways.[34] As a result, the ways in which racial/ethnic categories are linked to health outcomes are not stable and require further investigation. The use of ethnic categories is also unreliable because their constructions are historically and politically contingent. For instance, the term *Hispanic* was coined, and politically motivated, during the Richard Nixon administration to reclassify Mexicans, Puerto Ricans, and Cubans, who were previously considered "white."[35] Consequently, *Hispanic* became integrated into population demographics and scientific research. In addition, the term *Caucasian* represents a much older form of racial classification from eighteenth-century Europe, which situated "Caucasians" (including white Europeans) at the top of a hierarchy of classification closer to "God's original creation" and other "races" at the bottom, which included "Yellow, Red, and Black" people from different geographic regions.[36] The contemporary use of the term *Caucasian* reminds us that racist systems of classifications remain relevant, even if their racist origins are obscured.[37]

Social scientists and epidemiologists support the inclusion of ethnic diversity in clinical research "because having diverse research participants can improve the generalizability of medicine [and that diverse representation] touches on issues of health equality and the elimination of disparities."[38] Yet the unfolding of diversity and inclusion efforts in scientific, medical, and public health is multifaceted. From one angle, there exists more information, recognition, and awareness of health disparities, particularly in relation to obesity and diabetes, which has brought with it some helpful medical interventions. From another angle, these public health efforts are never completely disentangled from racist, eugenic histories of surveillance and exploitation, which make Brown and Black bodies, particularly women's bodies, more vulnerable to state management. From yet another angle, it became too challenging to continue the exclusion of Brown and Black people in data sets that shape national health standards—such as the BMI standards—because many began to recognize how this exclusion empirically undermines the production of medically effective standards. The point is that including ethnically diverse people in evidence-based medicine is both necessary and problematic because of racist eugenic logics that undergird existing epistemic approaches.

The 1993 NIH mandate focused primarily on counting and including more racial/ethnically diverse groups, which often resulted in comparing health outcomes across race rather than examining how racism impacts health. Decades after its institutionalization, the inclusion of ethnic racial minorities in medical research has not significantly ameliorated health disparities, or what Shannon Sullivan calls "racist disparities."[39] In fact, disparities across race and class for particular health outcomes, like preterm birth, have increased.[40] By exposing these tensions and contradictory outcomes of diversity and inclusion policies, the solutions that become possible might better address the core issues related to racism and its impact on health.

## The Impact of Race/Ethnicity across PIs, Staff, and Participants

One limitation in the scholarship on diversity and inclusion in clinical trial research is that it focuses primarily on the ethnic/racial representation of participants and not necessarily on the ethnic/racial representation of those who design, lead, and implement the trials. The PI of the SmartStart trial was aware of this limitation. For instance, in our first interview the PI referred to scholarship, which claimed that the successful implementation of clinical trials is not affected by the ethnicity of the PI in relation to the target population. She explained that a white man or woman designing and leading a trial that targets, for instance, African American pregnant populations, would not affect the trial results.[41] Having recently moved from the East Coast, the PI, who identified as a white woman, was aware of the cultural differences between her and the Latinx immigrant groups she worked with on the West Coast.

Regardless of the ethnic/racial identity of the individual PIs, it is well established in the clinical research literature that the recruiting staff and the referring physicians must be connected to the targeted population.[42] That is, it is necessary to have ethnically diverse staff and physicians who work within the targeted communities, and the US trial was designed with this in mind. The PI recruited local staff to help build rapport with the community. Identifying as Mexican American and having bilingual skills emerged as important in my eligibility to become a staff member in the trial. The majority of the staff in the SmartStart trial identified as Latina and spoke Spanish, a key benefit to recruiting "Hispanic" populations.

To contextualize the complex racial politics that structure trial recruitment and implementation further, it is important to note some of the benefits that staff received. Since the West Coast site of the SmartStart trial was small and not affiliated with a research hospital, the PI had more authority and flexibility in hiring staff. She mostly hired recent undergraduates and invested in their training. The opportunity to work on the study provided the staff members with research experience for graduate school or other community health projects. Three of the women I worked with in 2013 were still working with the PI when I returned in 2019. Although their pay had not increased as much as it potentially would have in a different industry job, the working environment remained reliable and flexible. The recruitment phase of pregnancy trials reflects larger long-standing issues of labor, race, and politics in research.

Moreover, the strategic staffing decisions of the trial also impact the retention of reliable labor and available participants. Since completing this prenatal trial, the PI has received another large NIH grant to implement a diet and exercise intervention with women before they conceive. Many of the diverse women targeted for recruitment for the SmartStart trial are also a part of this new study and the follow-up study with the children of participants. In a recent interview with a long-term Latinx staff member at the SmartStart trial, she told me that she has met with some of the same women for over six years. Having consistent staff is key to long-term participant retention, especially when the trial involves staff members collecting bodily measurements and sometimes visiting the participants' homes year after year.

### Referrals: Incentivizing Recruitment

The politics of race and neoliberal health-care infrastructure are made clear by exploring the incentives in the recruitment process of the US trial. The US trial recruitment relied not only on stable diverse clinical trial staff, but also heavily depended on the referrals of eligible pregnant participants from clinics and physicians. The doctors, nurses, and staff at various prenatal or women's health clinics were key to promoting the trial by recommending it to their patients. In order to get their support, the PI had to build relationships with the medical community in the area. The structure and rhetoric around recruitment in the US pregnancy trial are similar to those of for-profit pharmaceutical marketing strategies. For instance, recruitment strategies drew on marketing-oriented questions such as: How can we make this product or intervention appealing to the target population? Who do we need "buy in" from in order to help sell [or recruit people]? These approaches also shaped the strategy of incentivizing participating prenatal clinics' "rewards."

These business approaches to recruitment were particularly salient in the US context, primarily due to the lack of standardized and universal health care. In the second half of the twentieth century, biomedicalization marked a major shift in the global health-care landscape, which made medical services and diagnostic products available for direct to consumer consumption.[43] As a result, new markets for global medical tourism emerged, and efforts to market to individual consumers proliferated throughout the industry. Scientific research that engages with health-care providers across private and public spaces also uses business and marketing strategies to incentivize providers to support trial recruitment. Unlike private companies, the marketing resources for publicly funded trials are minimal; therefore, the "rewards" or incentives for recruitment into the SmartStart trial took the form of low-cost lunches and educational programming.

The clinic that had referred the most pregnant people to the SmartStart trial was a community health clinic that primarily served low-income "Hispanic" populations without health insurance. As a "reward," we organized a "lunch and learn" event. I helped organize one of these events, which included Subway sandwiches for the clinic staff to eat while the PI gave an educational presentation. We arrived at 12:00 p.m., and the staff gathered around the food while the PI and I set up the projector for her presentation. The staff who attended were all women except for one man, and the majority were Spanish-speaking.

I heard the staff chatting casually about eating, losing weight, and issues regarding the clinic. The clinic director commented to her assistant that they needed more bilingual speakers because most of the doctors needed translators. The social worker at this community health clinic, who was also Latina, talked to another group of staff members about a recent patient who had come from Mexico, eight months pregnant, with no previous prenatal care and no Medicaid registration; people gasped, reacted in a surprised manner, and rolled their eyes, and exclaimed things like "What was she thinking?" "How irresponsible!"

Capturing these moments of *chisme*, gossip or side chatter, is a practice of racial improvisation. Comments, gestures, and reactions come up and then disappear, sometimes leaving no trace behind. By including this brief moment, my intention is not to spotlight one individual person's perspective but rather to situate how particular comments and reactions can reflect broader political ideologies. The comment on "how irresponsible" this new patient is places total responsibility for the fetus's health on the pregnant individual regardless of international and structural policies that affect the conditions of health and migration.[44] Late liberal ideologies are entangled with xenophobic sentiments that shape pregnant participants' recruitment environments. These sentiments are also circulated within Latinx communities and reflect hierarchies of immigration status.

The stakes of xenophobic sentiments that underlie immigrant narratives in the United States are heightened when they target pregnant immigrants. Their bodies are not just carrying one but potentially two "invaders," as the conservative news media claim. In the summer of 2019, the US president introduced a rule that would penalize immigrants for using social services.[45] The racist and anti-immigrant sentiments perpetuated by US administrations adversely impact pregnant women and their *US citizen* children. In this way, strict immigration policies can have intergenerational effects, not only through epigenetics but also through the inheritance of immigrant status stigma. As Nicol Valdez's work outlines, undocumented parents already avoid engaging with state bureaucracies.[46] Strict identification policies like the recent "real ID," which requires extreme amounts of documentation to prove residence, make it impossible for mixed status families to seek out and receive services for their US citizen children.

In friction with the dominant xenophobic and racist rhetoric that frames immigrant communities as "stealing" jobs, resources, and services from the imagined white normative citizen, I emphasize that the US publicly funded clinical trials *need* these participants to volunteer their pregnant bodies and subject themselves to more surveillance and intervention. The scientific information that is produced from their bodies, data, and labor benefits the broader public.[47]

Once the crowd had settled down, the PI introduced her talk, which focused on successful strategies to maintain weight loss. The presentation reviewed a large survey that followed people through their weight loss journeys (while clinic staff ate the food we brought them). In examining the survey data, the PI found key strategies to maintain weight loss, including physical activity for an hour a day, low-calorie and low-fat diets, frequent self-monitoring, standard meals with no variability, routine breakfast, and water consumption. The crowd was vocal about their opinions, some commented: "They are so disciplined!" "They are lying!" The one detail that caught everyone's attention was the finding that most people who maintained their weight loss also increased their alcohol consumption. One staff member joked, "Oh my food is gone, let me take a drink!"

The PI finished her presentation with a key point: "Meal replacements can help with weight loss long-term. The studies have been repeated over and over, and this is remarkable." She emphasized that she did not receive any money from SlimFast but confirmed she was a believer in the efficacy of using meal replacement shakes. The director of the clinic commented, "But they taste so bad!"

Many of the strategies that the PI pointed out in her presentation are the very same strategies she implements in the prenatal trial. The PI confirmed to me that she is not receiving any money for using a particular brand of meal replacement beverage. Private companies had approached her to use their brands, but she refused their offers, claiming that the trial was not testing meal replacement shakes. This decision also protects the trial from being associated with the development of commercial products. Meal replacement beverages for pregnant populations are a growing market. In the UK trial, Abbott Pharmaceuticals was recruiting nonpregnant adults (including trial staff) to test its glucose-regulating shakes. Another emerging market that draws on the same pregnant population is Weight

Watchers in the United States and Slimming World in the United Kingdom. Both companies have recently started marketing their weight control and nutrition programs to pregnant people.

Whereas the "lunch and learn" represents a low-cost way of incentivizing recruitment at a public clinic, the business of recruitment looks quite
different in private, for-profit pharmaceutical drug trials. Unlike healthy
nonpregnant adults who participate in risky phase I trials and receive
financial incentives, the pregnant participants targeted for recruitment
in the US prenatal trial are not paid a living wage. As I have mentioned,
in total, the participants receive up to $175 for the completion of assess-
ments, tests, and multiple hours of participation in the intervention.
The modest amount of money that the US trial gives to its participants
is based on the rule that, unlike private pharmaceutical trials, publicly
funded prenatal trials are not supposed to significantly incentivize participants through payment. In addition, the amount of financial incentive
is also framed in connection to the amount of risk that participants take
on, which partially explains the high payment for phase I participation.[48]

Existing scholarship shows that global capitalisms play a fundamental role in the exponential growth of pharmaceutical drug markets and
research development.[49] In *Medical Research for Hire*, Jill Fisher examines the expansion of clinical trials within the United States and the simultaneous growth of contracted private physician centers that implement
pharmaceutical trials.[50] The private health-care and research sector has
more money to pay physicians to recruit and implement trials, so much
so that private small-practice doctors are financially incentivized to help
recruit subjects for private pharmaceutical clinical trials. In addition, the
pharmaceutical companies have larger budgets to pay human subjects. In
Fisher's recently published book she found that socioeconomic disparities
shape the supply of healthy human volunteers, who are primarily Black,
Hispanic men who are unemployed and many of whom were previously
imprisoned.[51] Importantly, I emphasize that the structures and systems
that shape the availability of healthy Black and Brown volunteers, and the
financial incentives in for-profit drug development, are indexical of racial-
surveillance biocapitalism.

The privatization of health care and research, the lack of stable employment, and income inequality create conditions in which people are

motivated to participate in clinical trials. Increasingly, vulnerable populations are required to make so-called autonomous individual "choices"
from extremely unequitable conditions and options. For folks who are
terminally ill, trials may be more of an option than a risky experimentation, and for those who have no health care, participation in a clinical
trial supplements basic care.[52] The socioeconomic inequalities that shape
clinical trial participation cannot be disentangled from systemic forms
of racism, which influence unemployment, the risk of incarceration, and
health disparities. These dynamic political, economic, and racist milieus
also shape the motivations for pregnant participants, but in nuanced
ways. Distinctly, the recruitment politics of pregnancy trials reveals how
forms of risk, race, and fatness are mobilized in the name of preventing
illness and speculating on people's future health.

## UNITED KINGDOM, STANDUP PRENATAL TRIAL

### Recruitment Infrastructure

The UK prenatal trial was part of a consortium that included five participating sites across England and Scotland. The consortium was awarded
just over £2 million from the National Institute of Health Research in
the United Kingdom. The UK consortium was centralized in its structure and design. It had one main site, in central England, with one main
investigator, who managed all other trial sites with the help of collaborators and project coordinators. In addition, all participating sites applied
the same intervention. The majority of my fieldwork took place at the
main site of the consortium. This hospital was part of the National Health
Service (NHS) in the United Kingdom and handled more than two thousand births annually, providing a large pool of potential participants. The
teaching hospital facilitated the centralization of all recruitment, data collection, and intervention delivery.

In 1946 the United Kingdom approved the National Health Service
Act, which states that the NHS was intended to "secure improvement
in the physical and mental health of the people of England and Wales
and the prevention, diagnosis and treatment of illness."[53] Health services
were provided to the public free of charge, and the act reorganized the

administrative management of hospitals under a regional board of gov-
ernors. The act legally made provisions for welfare services for the sick,
unemployed, women in maternity, and people of old age. Within a year of
establishing the NHS, the British government also legislated new laws for
the professionalization of nurses and midwives.

Since its inception, however, the NHS has endured many changes and
budgetary cuts.[54] When I was in the field, a guiding framework for the
NHS was the integration of teaching, care, and research. The synergy
across these three areas was multivalent; it was intended to mediate budgetary cuts, and it was framed as an innovative intervention to enhance
the system. One of the underlying messages was to "do more with less," a
common trope in twenty-first-century health-care contexts with growing
income inequality. The economic constraints, along with the integrative
approach, shaped the structure and context of the StandUp trial's implementation, including the recruitment phase.

On the ground, the integration of research and prenatal care manifests
through the space and location of the trial's implementation. The data collection and intervention delivery took place in the clinical research facility
(CRF). The CRF was a large area located on the fourth floor of the hospital, which housed many different research studies. The StandUp trial was
just one of many ongoing RCTs in the CRF, but it was the only prenatal
lifestyle study. The CRF was a laboratory and an experimental environment. The floor looked like a maze of grey rooms where wet labs were
mixed with MRI, X-ray, and other technological equipment. The data collection visits with the research midwives took place in exam rooms that
had sterilizing equipment, needles, a blood pressure cuff, a bed, a scale,
and a stadiometer to measure height. There was also a computer for the
midwives and participants to enter survey data. The intervention sessions
also took place in similar exam rooms. In general, it was a difficult space to
navigate, with extra security and winding halls compared to the maternal
and infant health floor. Participants were often late to their appointments
because they got lost or could not find the exam rooms.

The description of the trial site and the floors on which the trial intervention was implemented serves as a way to illustrate the health-care
infrastructure that enabled recruitment in the United Kingdom. It also
shows how lifestyle pregnancy trials are incorporated into clinical and

laboratory settings. Even though the UK trial was testing a diet and exercise intervention, not a drug or risky medical intervention, it was implemented in the same spaces as other trials. No matter the kinds of trials or risks, all the participants in the CRF were moving through the same space and generating biobehavioral data points for collection, storage, and eventual analyses.

The trial approached people for recruitment during their first two prenatal visits at the hospital. It had access to all scheduled prenatal visits through a standardized NHS database that included medical history information as well as the weight and height of all patients. This particular aspect of NHS infrastructure was vital for trial recruitment at large teaching hospitals. The research assistants, at the main site of the UK trial, were responsible for initially approaching eligible participants. The research assistants would scan the NHS database looking for specific participants that met the BMI requirement. Then they would approach each person at their prenatal appointment on the eighth floor of the hospital. If the person agreed to give their contact information, a midwife researcher would call them and set up their first trial visit. However, at smaller participating trial sites within the StandUp consortium, recruitment was much more challenging.

The other participating sites around the United Kingdom did not have the same resources as the main site. At the participating sites based in smaller towns, midwife researchers were in change of recruitment, enrollment, and data collection. Midwife researchers at smaller sites also had to write down everything by hand and then type it into an available computer after their meetings with the participants. During my conversations with other research midwives, there was a sentiment that the staff at the main site had it "easier" or were spoiled because of the resources at the teaching hospital. Although there were clear differences in resources and capacity across the various sites within the broader consortium, the trial needed the smaller participating sites in order to reach the recruitment goal of over a thousand women.

On a typical day at the StandUp trial, I would take a bus to the hospital, walk past the nurses and doctors who gathered in front of the hospital with signs protesting the major NHS budget cuts, and walk through the large main entrance. The main entrance had a flower shop and a grocery store where most staff bought lunch, and it was usually busy with visitors, patients, and staff rushing around. I went toward the back of the hospital to get to the large elevators that often smelled of "hospital": a kind of stale mixture of expired food, damp sheets, and sterilizing chemicals. The elevators were notoriously busy and slow. Riding up and down them while shadowing the staff became a routine part of the day. The staff had offices on the tenth floor within the women's health department. All biosamples, including blood and urine, were stored on the tenth floor. The fetal medicine unit, where most recruitment took place, was located on the eighth floor. Data collection visits, along with intervention delivery, took place on the fourth floor in the CRF. Usually I went straight to the tenth floor; checked in with the midwives; and followed them to the fourth floor, then back up to the eighth floor to do blood or ultrasounds, then back down to the fourth floor to complete the visit; and then finished the day on the tenth floor. If I was following the research assistants, I checked in with them first and followed them to the eighth floor to recruit the participants who were scheduled for their first prenatal appointments. Staff meetings and phone calls with all participating sites took place on the tenth floor near the PI's main office.

*Incentives for Recruitment*

None of the participants in the UK trial received any money for enrolling in the trial. The one type of incentive I did observe was the 3-D ultrasound, a more sophisticated, high-definition image that is expensive and not included in standard prenatal care. Participants who completed their second trimester assessment tests received the 3-D ultrasound. It was an effective incentive for retention, though not explicitly for recruitment. However, there were financial incentives for the staff, PI, and collaborators implementing and designing the trial. One of the project coordinators explained to me that for every participant they recruited, the PI and collaborators received about £1,500 toward their next research budget. From my understanding, the recruitment efforts in the present afforded the scientists and staff money for more research in the future. Once I heard the number—£1,500—I wondered what impact that money could have on the participants' health if they were to receive that amount distributed over the course of pregnancy and postpartum through a cash transfer program. The financial incentive for the scientists and staff for clinical trial research based on recruitment, but not for the participants, was a unique difference in the UK NHS setting.

At a conference I attended on the politics and ethics of clinical trials in the United Kingdom, I discussed the notion of paying participants for their labor in clinical trials, as Cooper and Waldby make the case for in their book *Clinical Labor*.[55] The suggestion was not well received because paying people to participate in clinical trials, even a very small amount, was viewed as unethical in the United Kingdom. In addition, the trials in the UK benefited from the public sentiment that participating in trials that are integrated into the NHS is a way to "give back" to the NHS. The motivating conception is that donating one's body, time, or labor is socially acceptable and even socially responsible because the NHS provides care, albeit not as much care as in the 1970s, 1980s, or 1990s, but still some health care for all residents.

Another caveat to consider is that private industry collaborations with publicly funded NHS research were allowed and promoted. And the potential future earnings of these collaborations are neither accounted for nor transparently documented. These dynamics—the increasing privatization of scientific research aimed at medical or pharmaceutical markets and the growing partnerships between publicly funded trials and private business—create conditions in which people designing and collaborating at elite levels have more opportunity for financial gain than the people who altruistically volunteer their bodies, time, and labor.

## Why Do Pregnant People Enroll in the UK Trial?

A main reason pregnant people enrolled in the UK trial had to do with their access to a midwife. This reason for enrolling in pregnancy trials in the United Kingdom is well documented in British health literature and is related to the NHS budgetary cuts to pre- and postnatal care.[56] In the latter half of the twentieth century, the NHS offered consistent and reliable prenatal care that depended largely on midwives. Pregnant people would get to see the same midwife throughout their pregnancy, and they would also get a postpartum midwife for an extended period of time if needed.

Those memories still linger in the minds of the older midwives I interviewed in the United Kingdom. Currently, however, pregnant women see multiple midwives, they have fewer visits—unless classified as high risk—and they do not have as much access to postpartum midwives.

During the trial visits with the midwife researchers, participants often asked extra questions about their pregnancy and birth plans. To this end, the two main midwife researchers in charge of data collection for the StandUp trial organized each visit so that they would see the same participants consistently. The motivation to spend more time with a midwife, even if it that means enrolling in a prenatal trial, reflects the integration of health care and research in precarious economic contexts. The prenatal trials in this milieu supplement prenatal care and make pregnant bodies available for research. This integration is structurally supported by policy requiring midwife researchers in the implementation of all UK prenatal trials. In this way, the policy to hire research midwives for pregnancy trials reinforces the existing motivations of pregnant participants who seek more time with midwives. In addition, as I mentioned on the intersections of fatness, those who chose to participate in the UK trial stated that they were worried about gaining too much weight and wanted extra surveillance or guidance during pregnancy. Although the issue of fat stigma also prevented people from enrolling, those who did enroll were self-selected; I explore their particular experiences in the intervention phase of the UK trial in the next chapter.

## Race and Recruitment: Classification and Improvisation

In the UK, the StandUp trial targeted a "broad" ethnic population, and the trial employed two separate classifications or codes for ethnicity: a code from 1 to 10 and a four-category code "Asian, Black, Other, White."[57]

1. European
2. Indian
3. Pakistani
4. Bangladeshi
5. Afro-Caribbean
6. African

7. Middle-Eastern
8. Far East Asian
9. South East Asian
10. Unclassified

In the code numbers 1–10, each number represented a different geo-graphical region that stood in for ethnicities. This list was posted on the office walls to help the staff remember the classifications, since they var-ied across different UK trials and over time. The statistician from the UK trial explained that the ethnicity code numbers 1–10 in the preceding list were designed by a different UK trial, which focused on birth weight and ethnicity. Since the StandUp trial was also collecting birth weight data, it included the ethnicity code 1–10 from the other trial in order to translate and share the data across both studies.

In addition, the StandUp trial applied a four-category code that was similar to and adapted from the categories used in the most recent UK census. Since the StandUp trial is a national trial, it is intended to reflect national demographics.[58] Using a broader set of categories like "Asian, Black, White, Other" that draw from the UK census categories would enable a national comparison of the results. Making trial results gener-alizable allows for the outcomes of the trials to be extended or applied to larger national populations, in this case that of the United Kingdom. Thus, the UK trial used two separate classification systems to ethnically/ racially organize participants. In my observations with recruitment and intervention implementation, I only witnessed the implementation of the ethnicity code 1-10. However, in the questionnaires and surveys that the participants filled out on their own, the race/ethnicity code included just four categories adapted from the UK census.

The classification list from the StandUp clinical trial in the United Kingdom intended to categorize ethnicities for the production of scientific knowledge. Yet the codes produce something more. In its application, I found the code to be a rich source for what I describe as the *improvisation of race*. The ethnicity code stimulated reactions, responses, and negotiations that helped bring different aspects of race into existence. In addition, the codes themselves are improvised. The rationale for applying certain clas-sification systems is fluid; they change over time, and they are mutable and hard to trace. For example, in 2019 I followed up with the statistician from the StandUP trial, who told me that the ethnicity code 1–10 was already "outdated" and no longer in use. It is not that those populations or people no longer exist, but that the classifications are outdated in relation to relevant demographics based on current research needs. Racialized target populations and the codes used to classify them in trials are mutable. The expiration of codes sheds light on the mercuriality of race in clinical practice.

Another difference between the ethnicity code 1–10 and the four-category code was that the latter, derived from the census, included "White," whereas the former included only the term "European." When I observed the collection of surveys that asked "What is your ethnic origin?," participants would respond as White British and not European. The midwives would have to make the adjustment in the process of inputting the data. They would say, "Here we use European and not White British." During the initial in-person data collection visits, the midwives modified the participants' responses to fit into the 1–10 code; however, participants could choose to identify as "White" and not "European" on the written surveys and questionnaires.

The use of multiple ethnicity codes in the StandUP trial shows how ethnicity codes can be adapted from a variety of places and for different reasons and are temporally unstable; this is the improvisation of race. The stakes related to how participants fill out and engage with the different codes are directly related to how these codes are used in the development of health policy and racial ordering. The improvisation of race on behalf of the participants, such as choosing between numbers 1–10 and "Asian, White, Black, Other," carries impacts for the creation of future ethnicity codes to classify target populations.

### "What Is Your Ethnic Classification?"

The first time I saw someone use the ethnicity code 1–10 was at a recruitment appointment when I was shadowing a research assistant, Jim, the only man working on the trial. Jim did not describe the code in the same way in which it was written. In a recruitment pitch with a potential trial participant, Jim used the term *ethnic classification*, which corresponded to a number on the recruitment form:

JIM: We are trying to improve women's health during pregnancy and reduce risks associated with GDM (gestational diabetes) and BMI (Body Mass Index). You are randomized into control or intervention [groups]. Control group meets with research midwives three times and has a GDM test at twenty-seven weeks. The other three visits deal with general health. Both groups get the GDM test. The intervention has three visits and works with a health trainer to make lifestyle changes. This is not a diet. Can I get your contact information? Any questions?

PREGNANT WOMAN: What is a health trainer, is it like a fitness guy?

JIM: No, just advice, to improve outcomes for pregnancy for all. What is your ethnic classification? (shows them the recruitment form). Circle a number 1–10. This part (pointing to the consent to access medical records form) is voluntary, if you choose not to participate in the study we would like to ask for your initials, post codes, outcome data.

Whereas Jim used the numerical code because it was labeled on the recruitment form and required the person to mark a number, the actual numerical categories were not referenced in the same way by all the staff members. For example, when I observed data collection during the in-person enrollment visits between pregnant participants and the midwife researchers (MR), they asked women how they identified, and then they would fit their answers into the categories 1–10 on their own. In my observations of these visits I found the responses varied:

MR: What is your ethnic origin?

PARTICIPANT A: Ghanaian and Italian.

MR: Brilliant, great food! (the MR silently enters a code for ethnicity into the computer).

MR: Ethnic origin?

PARTICIPANT B: Portuguese, and baby's dad is American and European.

MR: Putting that down as unclassified (which is number 10 in the ethnicity codes list).

MR: Ethnic origin?

PARTICIPANT C:   West African.

        MR:   We've got African or Afro-Caribbean.

PARTICIPANT C:   Okay, African.

There were two different midwife researchers enrolling the participants and collecting the data. They each had different ways of modifying the actual responses to fit the numerical ethnicity codes required for data collection. In the first response, it is not clear what the MR chose for participant A. Responses with multiple ethnic or national origins such as Chilean, Spanish, Portuguese, and American are organized into the category labeled "other." I observed this process repeatedly, and it was a routinized practice of the MRs to fit participants into the existing categories.

In the last example, the MR does not use the numerical reference to African (6) or Afro-Caribbean (5) but instead the actual terms from the 1–10 code. She tells the participant in a matter-of-fact way "we've got" certain options, and the participant has to adjust her ethnic identity to fit into the existing categories; this is a form of racial improvisation. I recorded fleeting moments like these out of a methodological commitment to capture the mundane material of clinical trial encounters. These moments show how staff and participants negotiate categories to improvise responses for the data collection in real time. The broader point here is that racial categories are imagined to be stable, but in practice the application of racial categories is messy and improvisational.

The results from clinical trials that document race and ethnicity (and most are required to) are often used to make comparisons, develop insights, and draw connections that link such categorizations to a "risky" behavior or to a kind of medical predisposition.[59] For instance, in the context of gestational diabetes, a form of diabetes during pregnancy, its prevalence in the United Kingdom is associated with "women of Asian and South Asian descent."[60] However, what counts as "Asian or South Asian" in the data collection phases of recruitment, in-person visits, and trial surveys is different partly due to the multiple codes used in one trial and the different ways in which participants improvise their responses across different ethnic codes. In addition, the "Asian" category organized in the United Kingdom is derived from a local population in central

England. The act of swiftly picking number 1 or 2 on a recruitment form enables the improvisation of race in a clinical trial setting and during data collection.

Racializing populations through improvised classification systems also influenced how certain populations were targeted for recruitment. A lack of comprehension about religion and food, for instance, created environments that were not very welcoming to certain demographics. As a result, groups that were classified as non-European were harder to recruit than "White British" participants. The trial also failed to anticipate the challenges to recruitment posed by (non-Christian) religious holidays or food cultures. One main challenge was recruiting women from Muslim communities. During my longer research trip in 2014, I was there during the summer months coinciding with Ramadan. The trial did not adjust to the recruitment challenges during the months of Ramadan, which impacted the diversity of the sample and the overall numbers of enrollment—not significantly, but it was mentioned briefly by one staff member chatting in the hallway. While pregnant women may be exempt from fasting during Ramadan at different stages of their pregnancy, there are still some who chose to fast during the day and would not be able to participate in a prenatal nutrition trial during this time. However, the issue of accommodating non-Christian religions and related dietary practices was not directly addressed in the recruitment meetings.

Religion did emerge as a way to characterize certain ethnic groups, for instance, for data collection purposes in the trial. In conversations about recruitment in staff meetings and recruitment encounters, research staff would at times describe the "South East Asian" category, labeled as ethnicity code 9, as "South East Asian Muslim Women." Not only were the categories of "South East Asian women" and "South East Asian Muslim women" used interchangeably, but the word "Muslim" was also referenced seemingly out of nowhere. The trial ethnicity code did not include religions. To add to the confusion, any or all of the other groups in the code 1–10 could also have people who identify as Muslim.

The "South East Asian" group (number 9) came up during recruitment meetings because the trial had a hard time recruiting women who identified as number 9 despite having a site in an area that has a high Southeast Asian population. One staff member stated that it was difficult to recruit this group because they were "a tight knit community." This was a passing comment, and the rest of the staff did not respond to it. In a different meeting, another staff member mentioned that the study had not enrolled any "Asian" women. This assessment of course depends on how the categories are counted. From the staffs' comments it was not clear who counted as number 8, Far East Asian, number 9, South East Asian, or just "Asian" in general. No serious or formal conversations were had about attending to the Southeast Asian and or the Muslim community differently for recruitment.

In addition, the trial did not customize or adapt the nutritional interventions for different ethnic foods, which may have impacted their ability to attract and retain women from immigrant and racially/ethnically diverse communities. As I examine in the next chapter, everyone in the intervention group was encouraged to eat a "Mediterranean" diet. This proved challenging for families that were not accustomed to eating, buying, or making "Mediterranean" foods.[61] For instance, a participant in the UK trial who was first-generation British with family from Senegal explained that she ate rice, okra, and palm oil as main staples of her diet. The staff member delivering the intervention noted that the participant needed to swap these foods out for other kinds of "healthier" foods. Such framings around healthy or unhealthy foods are not neutral or objective, but enveloped in racialized and colonial relations of power.[62]

Here, racial improvisation is methodologically and conceptually important for understanding how moments, gestures, and comments throughout clinical trial implementation shape emergent experiences of race/racism. The application of racialized categories in recruitment necessitated improvisation on the ground because the categories did not fit participants' experiences or identity. Moreover, different attention to and awareness of language, immigration status, food, and religion all impacted which communities were successfully targeted and recruited, and which had better retention and intervention compliance. Taken-for-granted assumptions about diverse populations became barriers to recruiting populations that have systematically experienced racism, violence, or exclusion from public health-care services and medical treatment. The next chapter shows how these same assumptions shaped the effectiveness of the intervention delivery.

The recruitment of ethnically diverse pregnant participants is significant for understanding another layer of racialization that occurs in prenatal trials: the improvisation of race during pregnancy has consequences also for the future children born into the study. The race or ethnicity of the children will be chosen/improvised by the pregnant participants, who are in charge of filling out the survey data forms for the newborns. In this context, the employment of ethnicity codes illuminates how epigenetic ideas are translated in and through racialized pregnant bodies.

The analysis around race, which foregrounds improvisation, throws into relief the messy, nonstandard, and imprecise ways in which race is enacted by evidence-based science. Including race as a variable for targeting and recruiting participants in clinical research does not inherently address or examine processes of racism or issues of equity. Not only do systems of classification require more rigorous attention, but the underlying motivation and focus on including and classifying people draws attention away from how white supremacy and racism shapes health outcomes for vulnerable populations.

.     .     .     .     .

This is one of the first ethnographically grounded explorations of clinical trial recruitment across the United States and United Kingdom, specifically one that focuses on pregnant populations. The firsthand understanding of recruitment that I gained by working as a staff member in the US trial provides context for how pregnancy trials are implemented. The mundane processes, practices, and forms that make up recruitment in US and UK trials reveal relations of power that are overlooked when clinical trials are only evaluated by their findings and not by implementation. With such an overwhelming focus on trial findings, scientific publications rarely provide an inside look into how trials are made, and more importantly, into the political, cultural, and economic contexts that shape human subject recruitment. Whereas most of the social scientific literature has focused on global pharmaceutical trials, I emphasize that contemporary pregnancy trials in the Global North are an understudied site that reveals important processes of scientific knowledge production writ large, specifically aspects of recruitment that are connected to themes of race, risk, and fatness.

A close examination of the recruitment phase of pregnancy trials also illustrates the distinct aspects and common outcomes of US and UK health-care landscapes. Both the US and UK prenatal trials were implemented in socioeconomic climates characterized by disinvestment in social and public health-care safety nets. Such epistemic environments shaped who was targeted, the role of risk in justifying target populations, and people's motivations to enroll in the trials. The different private and public clinics from which the US trial recruited reflect the uneven care and surveillance people receive. In the United Kingdom, the NHS played a unique role in bolstering recruitment efforts. It emphasized a "synergy" across clinical research, teaching, and care to supplement gaps in health care. In this way, participants were motivated to enroll in the prenatal trial in order to receive extra care from the midwives, indexical of late liberal and capitalist contexts at work.

Regardless of the different health-care landscapes, both the US and UK trials experienced unique challenges to recruiting pregnant people with a BMI of 25 or higher. Fat phobia both motivated people to take part in and dissuaded them from participating in these prenatal trials. Those who were worried about gaining too much weight were motivated to participate in the trial, and those who did not participate were worried about being fat-shamed during pregnancy. Next I explore how participants engaged with the intervention itself and how this impacted body image and anxieties around controlling weight.

# 4   Pregnant Narratives

EXPERIENCING LIFESTYLE INTERVENTIONS

StandUp pregnancy trial, United Kingdom, 2014:

PARTICIPANT:  In my last pregnancy, I was huge, and I got GDM [gestational diabetes mellitus]. It was a horrible pregnancy, I had to inject insulin seven times per day. I wish I had something like this [trial] last time. I've lost half my body weight since my last pregnancy. Because I lost so much weight I hope this time will be better, compared to what I was. I'm afraid to put on some weight. I've been in this mind set of lose, lose, lose it's hard to adjust.

MIDWIFE:  You've done fantastic! Your kids must be happy.

[*Participant steps onto the scale.*]

PARTICIPANT:  O.K. what is the damage? Oh, I gained!
It's O.K. It's O.K. It's O.K.

MIDWIFE:  Regarding your weight, has anyone else said you don't weigh enough?

PARTICIPANT:  Someone said I was anorexic—[*laughs*]—it's just because I'm afraid of gaining too much weight, which you can't understand if you've never been so big you could die.

• • • • •

Imagine feeling this scared to gain weight during your pregnancy.

Pregnant participants in the UK trial had their weight documented and tracked. Most people in the National Health Service (NHS) prenatal care system do not have their weight measured at each prenatal care visit because current policy claims that monitoring gestational weight gain (GWG) only provokes more anxiety and does not significantly improve pregnancy outcomes.[1] Stepping on the scale generated a variety of reactions from participants. A majority of the participants I observed would get on the scale and react in a surprised or concerned manner. Similar to the participant's reaction above, participants would gasp and say "Oh my god!"[2] Others would ask for their weight from the previous data collection visit, and the midwife would always share that information. In one case, a woman calculated that she had gained two stone (twenty-eight pounds) between her first visit at fifteen weeks' gestation and her second visit at twenty-six weeks' gestation. This weight gain made her visibly worried. She then asked the midwife: "Is this okay? Should I be gaining this much? You would tell me if I am gaining too much, right?" The midwife responded as she responded to many others who were worried about their weight gain: "There are no recommendations for how much weight you should gain. The baby is growing, you are alright." The response soothed the participant for the moment.

Understanding the fear and anxiety around weight gain during pregnancy requires some historical contextualization of the concept of "obesity" as an illness and its relationship to body image. On a population scale, a great shift called the *nutritional transfer* occurred after World War II, which is characterized by the industrialization of food production and distribution of processed, cheap foods globally. This illustrates our current "obesogenic environment," a term coined in 1999.[3] This was a global movement shaped directly by economic policies of free trade, large-scale production, marketization, and consumption, not individual choices. As Gálvez proposes in her book *Eating NAFTA*, "The massive proliferation of diet-related illness [is] a kind of structural violence—a result of policy decisions and priorities."[4] In addition, existing research shows that the intersection of racism and poverty is a stronger force that influences health more than weight gain alone.[5]

Nevertheless, contemporary framings of obesity claim that the problem and the solution lie completely in your hands and what you decide to put in your mouth.[6] Lifestyle interventions that focus on changing individual bodies and behaviors selectively ignore the longer history and structural understanding of obesogenetic environments.[7] The practices and policies of disinvesting in welfare programs, the focus on the future over the present, and the hyperfocus on individuals are indicative of the epistemic environments that shape the imagined problems and solutions of future health.

Another compounding aspect that shapes the anxieties and fears of pregnant participants in the trials is body image.[8] Body image and beauty standards intersect with issues of racism, and fat phobia. To be thin (and white) is a signal of both beauty and health in dominant cultures.[9] Body image and representations of cis-women's bodies in particular are part of a larger cultural hegemony that is based on white supremacy, heteronormativity, and racial capitalism.[10] Thinness is also associated with being a "good" citizen, which is challenging to become in a neoliberal and racist society that makes it impossible for poor people of color to reach the expectations of "good" citizens.[11] The conflation of thinness with health also obscures other health issues that result from oppressive and unattainable beauty standards. Currently, the rates of disordered eating that develop into illnesses like anorexia, bulimia, orthorexia, and others are higher than ever before.[12] The degradation of mental health due to fat phobic, racist cultures is not accounted for in national health-care costs.[13] Arguably, the pressure to be thin and have a body that represents the ideals of a proper neoliberal society also contributes to adverse health outcomes.

To be sure, there are serious health issues surrounding metabolic diseases that are complexly related to food, eating, environments, stress, history, family, and trauma.[14] However, there is no scientific consensus on the metabolic etiology of obesity. The point I emphasize is that lifestyle interventions that focus solely on weight control are not changing the health disparities in diabetes and heart disease in the United States or United Kingdom. Lifestyle prenatal interventions aimed at weight or glycemic control have little to no significant impact on pregnancy and fetal/infant health outcomes.[15]

The focus on pregnancy as a critical period for epigenetic modifications has increased the stakes of lifestyle interventions with overweight and obese pregnant people, in an attempt to control future health. However, scientific discourses within postgenomics/epigenetics and developmental origins of health and disease (DOHaD) have all but ignored pregnant people's experiences. In general, and across disciplines, there is a dearth of information on people's experiences with eating and weight management during pregnancy. Unlike the biobehavioral data that are systematically collected and speculatively valued (see chapter 6), participants' experiences are not systematically collected and analyzed in these large-scale pregnancy trials.

In response to the significant gap in the existing literature, this chapter centers overweight and diverse pregnant people's experiences. This is one of the first ethnographic accounts of pregnant people's experiences in an ongoing contemporary lifestyle trial. Drawing on the framework of the politics of postgenomic reproduction, this chapter makes a case for why and how pregnant narratives are important for understanding the implementation of lifestyle pregnancy trials. It also gives a voice to the people whose bodies are the sites of knowledge production, and in doing so reveals how disconnected the scientific findings and discourses are from the embodied experiences.

I draw from interviews with staff members and my participant observation notes of intervention sessions to characterize the dialogue between health trainers and participants. Although I did not have permission to interview individual participants in the trials, I had access to observe all intervention sessions, take notes on intervention delivery visits in the UK trial, and interview all staff including the midwives and health trainers who worked closely with the pregnant participants. I was also able to interview two former participants who had completed the trial and returned to work as staff. Their unique experience as both staff and participants helped further contextualize participants' experiences.

These pregnant narratives reflect emotionally and socially complex relationships to their bodies, fatness, eating, and diets. The themes of control and responsibility are employed in framing the solutions and problems of obesity; are embodied in the pregnant participants' experiences; and are also enveloped in epigenetic framings of the maternal body, which I discuss at the end of this chapter. Discourses of control and individual responsibility remain consistent across reproductive politics of the twentieth century and postgenomic reproductive politics of the twenty-first century. The pregnant narratives explored here are important to consider not only for imagining future health, but also for creating health-care interventions that are meaningful in the present.

## THIS IS NOT A DIET

Going on a "diet" to lose weight during pregnancy is not culturally acceptable in the United States or United Kingdom. In practice, however, changing how you eat during pregnancy functions like a diet whether it takes place in a clinical trial or a commercial weight loss program. Current maternal health policies circumvent terms like *weight loss* or *diet* by focusing on a rate of weight gain. The Institute of Medicine (IOM) in the United States recommends that people with a body mass index (BMI) of 30 or more should not gain any weight during pregnancy, or only gain half a pound or less during each week of pregnancy, which amounts to about twenty pounds in a forty-week pregnancy. To maintain a constant rate of weight gain (or not gain any weight at all) while the fetus continues to grow requires some net weight loss. The United States provides clear and strict gestational weight gain recommendations, and the US trial intervention draws on them to implement its calorie and weight control intervention. In contrast, the United Kingdom does not offer any maternal weight guidelines, but the trial is based on glycemic control, and participants often conflate this approach as a weight loss "diet" intervention.

In an interview with the dietician who helped design the UK intervention, she asked me about the US trial, and I described the intervention and how it was guided by the IOM recommendations. Her reaction was stern: "That intervention would be unacceptable in the UK."[16] She went on to say that people in the United Kingdom do not recommend monitoring weight during pregnancy. The dietician explained to me that the intervention draws on "control theory and social cognitive theory."[17] The protocol manual for the UK trial refers to the intervention as a "behavioral intervention designed to improve glycemic control."[18] The glycemic index (GI) is a measure of how saturated fats, carbohydrates, and sugar affect blood glucose levels.[19] Foods that are high on the GI are supposed to be avoided or controlled. The semantics around controlling grams of sugar compared to calories obscures the fact that both approaches are based on control and are only epistemologically different to the scientists, not the participants.[20]

One of the most common and curious statements that I heard during my observations of the intervention at the UK trial was: "This is not a diet." In my analysis of the interventions, I found that clinicians and participants used the term *diet* in two ways: diet as a food regimen (specifically an intervention), and diet as a weight loss program involving food restriction. For instance, calories were often referenced in relation to dieting, and some participants conflated counting teaspoons of sugar with counting calories. Health trainers repeatedly reminded the participants that counting or controlling grams of sugar or fat was fundamentally different than counting calories.

The health trainer's handbook provided specific guidelines and justifications for how to deliver the intervention. It was written by the principal investigator (PI), social scientists, the dietician I mentioned previously, and a group of postdoctoral students.[21] The key message for the participants is that the intervention focuses on controlling sugar levels.[22] The handbook also explains that the health trainers should not include any other information from "other programmes" such as other dietary programs or individual understanding of food and diets. The UK trial is intended to test the effectiveness of this particular intervention on obese, ethnically diverse pregnant women: "It's therefore really important that you follow the content and structure so we know the information given to all the women is the same – otherwise we won't know what has, and hasn't, worked."[23] This was a similar concern in the US trial, where I was reminded that "I must deliver the intervention the same to everyone," no special treatment or attention to processing the participants' lives, as I illustrate in the next chapter. The conceptual boundaries between prenatal lifestyle interventions and dietary programs for weight loss are important to maintain for the sake of scientific research and evidence-based medicine. The handbook stated that "the intervention is *not a diet and will not involve calorie counting* [. . .] we suggest healthier alternatives rather than just telling people to avoid certain foods."[24] From the perspective of the trial staff and PI, the intervention is "not a diet," and therefore it is a better option for pregnant women.

From the perspective of the participants, programs like Weight Watchers and Slimming World were very similar to the prenatal interventions. In one of my observations of a trial visit, the participant was visibly disappointed because she was randomized into the control group. She said that she really wanted to work with the health trainer. She had gained so much weight in her last pregnancy that she wanted to do things differently during this pregnancy. When she found out she was not going to be in the intervention group, she decided to "do her own thing" and joined Slimming World.[25] She said that it does not promote "losing weight" during pregnancy, but it does weekly "weigh-ins" every Saturday.[26] In terms of her diet, she said that Slimming World encourages eating fresh fruit and vegetables. She also started exercising during this pregnancy, which she had not done in her two previous pregnancies. In this case, the participant decided to pay for a program that monitors and surveys her eating and exercise during her pregnancy, while also volunteering in a prenatal trial. What she did outside of the trial did not get documented in the trial, even if her activities might muddle the trial design, which required standard bodies and behaviors in the control versus intervention groups.

In another visit, a woman in the intervention group commented that she had been doing Weight Watchers before her pregnancy.[27] She said that the intervention was similar to the Weight Watchers program, except Weight Watchers focuses on "how much you're taking in through calorie counting and they monitor weight." The differences between the UK intervention and Weight Watchers program only became clear to her in the intervention. For instance, for the trial's intervention, rice cakes were not a healthy snack because they were high on the GI, but for Weight Watchers it was a healthy snack option. The same participant noted: "Oh no, I didn't know rice cakes were not health swaps! I ate like ten this week! Rice cakes, really? Rice cakes are in every Weight Watchers diet, but I guess they aren't paying attention to the sugar." In this way, regardless of whether the intervention aimed to exclude outside information from the prenatal trial, the intervention was not applied on a blank slate and could not control for what participants in the standard group decided to do outside the trial. Pregnant participants came in with their own experiences and knowledge of diets, and for them the intervention was like any other diet.

Health trainers had to straddle the line between meeting the trial's expectations and the participants' expectations. The health trainers could not offer any advice or recommendations on how much weight the participants should gain, and they reminded the participants that the main goal of the intervention was not to lose weight. Yet the health trainers also understood the participants' desire to monitor their weight and how past experiences with dietary programs shaped their compliance with the trial's intervention. One of the health trainers explained to me: "Trying to reduce sugar intake is a diabetes thing, but [they] realize, if I stop eating the biscuit, I'll lose some weight too."[28] The health trainers carefully balanced two realities: the nutritional intervention may help participants manage their weight during pregnancy, and yet it was not intended for weight loss.

Each actor in the trial framed and justified the intervention in distinct ways: The staff and health trainers in the trial had to stick to the party line that the intervention was not a diet, the participants applied their own desires and agency in their compliance with the intervention, and the PI and her scientific collaborators understood the intervention in a conceptually different way altogether. In the next chapter I explore further how the PIs in each distinct trial used epigenetics to justify the importance of nutritional interventions during pregnancy. At the UK trial, the PI was the only one who used epigenetics to explain the risk of high blood glucose levels during pregnancy for fetal development. The staff and participants never used the term *epigenetics* in trial implementation. The selective reference to epigenetics is reflective of the dynamic that epigenetics is theorized in elite spaces, tested on pregnant bodies, but not necessarily communicated with pregnant persons in a formal medical manner. Regardless of the different frameworks at play in the trial, everyone involved agreed on one aspect: pregnant women are responsible for and in control of changing their diets to protect fetal development.

## MANAGING WEIGHT: BEFORE, DURING, AND AFTER PREGNANCY

I gained more insight into participants' desire to manage their weights by interviewing two people who had unique insider experiences at the UK trial. Candice and Sheryl were staff members in the StandUp trial when I interviewed them; prior to that they had been enrolled in the trial as participants, and both were randomized into the intervention group. Candice was a well-regarded, lead, research midwife and had worked on the UK trial since the beginning; she noted that she had recruited the first participant to enroll in the trial. She became pregnant during the course of the trial and decided to enroll in it as a participant; after the birth of her child she returned to the trial and continued working as a lead research midwife on the trial. She was also the only staff member (in the whole trial), and of a high rank, who identified as both biracial and gay. Following the NHS structure, the UK trial had a hierarchical organization of ranks or "bands"; band 9 (a high rank) was a PI and band 3 (low rank) was a health trainer.[29]

Candice had significant experience with dieting and losing weight. In an interview, she told me about her struggles and victories with weight loss before getting pregnant. She even shared a "before and after" picture, which is documented on a national UK television program that applied hypnotherapy to subjects. She lost over a hundred kilos in the process. Losing more weight was not a key motivation for her to join the StandUp trial. Instead, Candice said she joined because she has a medical history of diabetes and heart disease, which made her feel like she was a high-risk pregnancy. As I mentioned in chapter 3, another factor that motivated Candice to participate was her loyalty to the trial and the NHS; she commented that she wanted to "give back" and help future women by volunteering in science. However, Candice made it clear that her past experiences with different weight loss methods influenced her perspective and experience in the intervention. She said that she already knew most of the nutritional material covered by the intervention, and that it did not necessarily impact her significantly.

Whereas Candice was based in central England, which is culturally referenced as "the South" and is viewed as a location of white privilege, Sheryl, a white Scottish woman from "the North," first came to the trial as a participant and later returned to the trial to work as a health interventionist.[30] In the UK context there exists a "North-South" divide, in which London is part of the "South" and is viewed as more privileged, elite, and as "ruling for the rest" of the United Kingdom.[31] These tensions heightened leading up to, during, and after Brexit. The North-South tensions are further complicated by racialized dynamics. The fact that Candice was the only gay and Black person in the higher ranks of the trial, and that she was from "the South" of England, was made more apparent in relation to the predominantly all-white staff situated in the North and South.

In an interview I did with Sheryl, she told me that she had decided to join the program, like most women do, in order to watch her weight. Sheryl has three small kids, is a trained nurse, and decided to work as a health trainer on the study. During her last pregnancy, Sheryl was recruited into the UK trial, and after her delivery she decided to apply for a job to work as a health trainer. She enjoyed working on the study, even though health trainers earned less than nurses in the NHS.

Within the context of the NHS labor hierarchy, Sheryl's decision to not work as a nurse, band 6, and instead work as a health trainer, band 3, stood out to her colleagues. However, in Sheryl's narration she explains that participating in the trial itself shaped her decision to return and work for the trial as a health trainer. In addition, her explanation for why she decided to participate in the trial reveals how she perceived the motivations of most trial participants. When I asked Sheryl why she joined the trial, she explained:

> Probably for like most people, not to gain too much weight. [. . .] I made the mistake during my first pregnancy to "eat for two" and I gained 2 stones and it was difficult to lose the weight because of returning to work, and balancing the kids, and it's not easy when you have a wee one at home. . . . 80% of the girls don't want to gain too much weight during their pregnancy, some will be conscious about GDM if they have a family member with it, . . . but 9 times out of 10 its usually just about the weight gain—not to repeat the same mistakes.[32]

Sheryl echoes the same concern and anxiety about weight that many of the pregnant participants had. The concern with gaining too much weight preoccupies women's minds before, during, and after pregnancy. An awareness of weight among staff and pregnant women reflects an existing issue with weight outside of the context of the trial. Regardless of how the trial frames the intervention, weight plays a significant role in its implementation.

Candice and Sheryl had different motivations for joining and working on the trial based on their individual experiences. Unlike most participants, Candice did not enroll in the trial to lose weight (because she had already lost a hundred kilos through hypnosis), but rather to receive more surveillance, or care, because she thought she was a "high-risk pregnancy" and she wanted to "give back' to the NHS. Her career as a health worker, along with her public weight loss experience, influenced her understanding of the trial's intervention. Sheryl, on the other hand, expressed a distinct motivation to join the trial specifically to lose weight. She believed in the intervention so much so that she decided to work as a health trainer on the trial instead of in her job as a nurse in the NHS.

As staff members and participants, Candice and Sheryl reflect a common concern around weight that existed among staff members and participants alike. During my first visit to the UK trial in 2012, the health trainers and staff discussed how working on the trial made them more aware of their own weight and diets. The staff members had internalized this concern with weight even though they were trained to tell the participants that the intervention was not a diet for weight loss. Staff members like the health trainers and research midwives, along with the participants, shared a wide range of weight loss strategies prior to and during their engagement with the UK intervention. The range of weight loss strategies that were discussed by staff and participants included bariatric surgery, liquid diets, and participation in programs like Weight Watchers and Slimming World (both of which recently allowed people to join during pregnancy).[33]

## PREGNANCY NARRATIVES

The intervention sessions involved a health trainer and a pregnant participant, usually one on one and in person; only a handful of times did I observe an intervention session done over the phone or an in-person session with more than one participant. The sessions lasted about one hour, during which the participants were given diet goals and physical activity goals to work on in between the sessions. All intervention sessions took place in exam rooms located in the Clinical Research Facility on the fourth floor of the hospital. Similar to the US interventions, the UK sessions involved intimate discussions about participants' bodies, families, work, anxieties, and motivations. The personal and intimate conversations that took place during the intervention appeared at odds with the sterile clinical laboratory setting of the exam rooms. Whereas chapter 5 explores the experiences of the pregnant participants in the US intervention through Iris's case, this chapter focuses only on the UK intervention. In the United Kingdom I was able to observe many different intervention session deliveries because I was not a staff member, and the hospital site I observed recruited the largest number of participants.

During my first visit to the UK trial in 2012, there were two staff members dedicated to handling the intervention delivery in the central England site. When I returned for longer in 2014, the main health trainer had moved to another hospital. The one remaining health trainer, Diana, handled all the participants in the intervention arm of the trial at the main teaching hospital. Diana had a background in holistic medicine and acupuncture. She was in her late forties and was the first generation of her family to be born in England after her parents migrated from Jamaica. She was well educated, with various certifications in health promotion. Diana was the only Black and ethnically/racially diverse person working as a health trainer at the main trial site.

The protocol manual of the StandUp trial states that the intervention must be delivered between twenty and twenty-eight weeks' gestation. In this window of time the intervention required participants to meet with health trainers each week for about eight visits. However, in my observations it was rare for any one participant to meet eight times. Technically, anyone who came to at least one session with the health trainer was included in the data analysis as a participating member of the intervention group. The biggest challenge to the intervention delivery was confirming and scheduling visits. Most women had more than one child and full-time jobs, and it was difficult to take time off during work. However, one significant difference between the US intervention delivery and the UK intervention was that the NHS recognized the UK trial as a form of prenatal care. This recognition by the NHS provided the necessary justification to take time off work to complete trial visits. In the United States, if participants were employed, they had to squeeze in time to do the intervention before or after work or during lunch hours.

## Shaina: Not a Diet but Education

During a first visit with Shaina, a trial participant, Diana asks her to explain her regular food habits. Shaina is in her late twenties and identifies as Afro-Caribbean. Much of her family still lives in Jamaica while Shaina is studying accounting at a university in the United Kingdom. She is also parenting a three-year-old as a single mother. Shaina describes her diet through her daily activities, such as drinking "fizzy drinks" or soda, eating too many chips, and "nibbling fruit at Uni [university]."[34] Diana responds by reminding Shaina of the intervention's benefits, including physical fitness for labor and a healthy pregnancy. As the handbook stated, Diana needs to make sure the pregnant participants understand that the intervention is not a diet. She says that the intervention focuses on "maintaining sugar levels" as a way to ensure a health pregnancy. Diana makes this clear throughout the intervention sessions.

In the same meeting with Shaina, Diana goes over the amount of sugar that is in a regular soda, and again she reminds Shaina that the intervention is "not a diet."

> DIANA:   1 coke = 7 tsp of sugar! Visually it's quite a lot.
>          Once you become aware you can make swaps.
> SHAINA:  Is that the same as calories?
> DIANA:   Well, we aren't counting calories, this isn't a diet.
> SHAINA:  Well, after baby I'm planning on going on a diet.
> DIANA:   Rather than diets this is making educated choices that can help long
>          term.

In response to whether counting teaspoons of sugar is the same as counting calories, Diana reminds the participant that the intervention is not a diet, but a way to make "educated choices" in the long term. This distinction may not resonate with the participant since she is still planning on "going on a diet" after the baby is born. From this particular session with Shaina, it is not clear whether she views this program as an educational intervention, in the way Diana believes it to be, or a kind of diet aimed at weight management.

The participant's question about whether seven teaspoons of sugar are the same as calories reflects a gap between her understanding and the scientific framing that the trial is presenting. The participant does not have a history or prior experience with counting calories or evaluating foods based on grams or teaspoons of sugar. During another part of the same meeting, Shaina was surprised to learn that white sugar and brown sugar are the same, or that one does not have more or less sugar content. In another instance, Shaina learned that the first ingredient on a label represents the ingredient that is used in larger proportion than the rest of the ingredients. Through these lessons, Shaina is being taught a particular way of reading and examining food labels. She is learning how to focus on controlling sugar or becoming aware of sugar content for the purposes of monitoring her blood glucose levels.

Since Shaina is interacting with this approach to food through glycemic control for the first time, her experience in processing and incorporating the intervention information is different from that of someone who has more experience with "counting" or reading labels. This is echoed in the remarks of some of the other participants. Diana indicated that on average the women of "African descent" do not have as much experience with dieting or counting calories as the white British women do. She said that as a result, the Afro-Caribbean or African women were more open to the intervention and less resistant because the information was new to them.

For Shaina the relationship between calories, sugar, and weight loss is one of similarity. However, for the aims of the trial, which Diana is trying to implement, there is a clear difference between counting calories and being aware or "educated" about the amount of sugar and saturated fat in food. To maintain the legitimacy of the trial, the distinction is important because the trial is part of science, not a fad diet. The distinction that is belabored by the staff seems to fade in importance for the participant, to whom counting sugar and counting calories are two sides of the same coin, which is still based on controlling diets.

### Mary: Learning to Behave Like a Proper European

Mary was a reliable participant. She came to each one of her sessions with Diana. Mary, a first-generation British citizen whose parents and family were all from Senegal, self-identified as West African and was studying for her master's degree in computer engineering. During the last intervention session, Diana asked Mary, "What are the main staple foods in Senegal?" Mary replied that they were rice, okra, and palm oil. Diana commented that most of those foods were high on the GI. She proposed that when Mary went to Senegal to visit her family, she would need to focus on portion control. Mary responded: "Portion, portion, portion, I do not want to share a house with you [Diana], too much portion. I just put a plate of food down and as long as you want to eat you just mix, eat, and tummy is full." Mary gave a huge smile, and both women started laughing.

The session continued, and Diana asked Mary, "What's been your biggest achievement and challenge?" Mary responded that her biggest achievement was how the intervention had "changed the way I eat, the way I think about food—I behave myself more." Again, she smiled. To address the second part of the question, Mary went on to say that one of her biggest challenges was being "aware of everything, like portions and liquid beverages. I'm not used to it, it will be hard because I've been eating this way for years."

Controlling one's portions was at first a foreign concept to Mary. The idea that one would measure a "serving" of food calculated by grams of sugar, carbohydrates, and fat is a Euro-American approach to food, eating, and sharing. As Mary mentioned here and in other conversations, she usually just put a big plate of different kinds of food in the middle of the table, and her family would take what they wanted. Encouraging the idea of portion control intervenes in a cultural and social way of relating and sharing food with others. The idea that the intervention made her "behave more" reflected the underlying notion that how she grew up eating in her Senegalese immigrant household was unruly, unhealthy, and even risky.

As Diana had mentioned to me earlier, the participants with no background in "dieting" were more open to complying with the intervention. An overlooked consequence of testing out nutritional plans in clinical trial settings is that they come across as scientifically and politically neutral. However, the idea that glycemic control is objectively healthier than a Senegalese diet reflects what Burnett calls *nutritional colonialism*. Burnett argues that ignoring the structural violence that constrains conditions of eating and well-being for Black women in Brazil and instead focusing on individual responsibility to eat "right" from a Western hegemonic framework represents nutritional colonialism.[35]

The UK intervention largely promoted a "Mediterranean" diet as the ideal. This culturally and economically inaccessible diet—itself an invention of US chemist Ancel Keys—illustrates how so-called treatments still operated within a racialized system of exclusion that set patients up for unattainable diet goals.[36] Tompkins argues in her book *Racial Indigestion* that eating Euro-American "local" diets "was a way to produce a moral body."[37] Scholarship in feminist and critical race studies of science illustrates that racism can creep into our food and bodies through the Trojan horse of "neutral" science and "healthy" interventions.[38] Disciplining or teaching participants to eat in a "healthy" way is substantively linked to forms of racialization and (neo)colonial politics.

### Donna: Liquid Food Diets, IVF, Risk, and Responsibility

In other intervention sessions, participants are explicit about their motivation to lose weight prior to getting pregnant and their experiences with other forms of dieting programs. In a first visit, a pregnant participant shared her experiences with other weight loss programs and related these to her desire to be "healthy" and lose weight for her pregnancy. Donna is a white British woman in her mid to late thirties.[39] She is a schoolteacher and had experienced five miscarriages before this pregnancy. Prior to this pregnancy she was told by her doctor that if she wanted to try in vitro fertilization (IVF), she would need to lose weight. Donna explained how she lost the weight prior to getting pregnant.

DONNA: Last year I lost 2.5 stones [around 37 pounds].
DIANA: What did you do?
DONNA: I did something controversial: I did liquid beverages. I replaced all food with liquid; it was hard at first but I had to do it, I was pushing 20 stones.
[My] BMI was too high for IVF, so I had to get my weight down to conceive, to have a baby.

In this exchange, Donna reflected on her motivation to do a "controversial" form of dieting. Interestingly, she used the term *controversial* to describe her perception of what is appropriate or not in the UK setting.

Donna submitted herself to the process of weight control and surveillance in an effort to reach her desire for conception.

It is clear that Donna was familiar with dieting, and in the next exchange Diana asked Donna about her regular diet.

DIANA:  So how is [your diet] now?

DONNA:  I get up, have cereal [and] milk, snack on some melon grape roll, and a fizzy drink.

[*She whispers when she says fizzy drink, as if it is a bad secret.*]

I don't like to eat sweet stuff, although I am heavier. I don't eat sweets, I like savory, my husband and I, we write down our foods, menu plan—I tend to worry when I get hungry. I get worried because I think I'm not just eating for myself. I don't want to get lightheaded. To be totally honest I need a sugar fix [*giggles*].

DIANA:  No judgment—So 100 ml of fizzy =2 teaspoons. So those beverages will have around seven teaspoons of sugar. What you experience your baby also experiences, [so you and your baby] will have a spike in sugar.

DONNA:  I think with me this is my fifth pregnancy attempt. I really want this, I don't want to do anything that would risk it.

In describing her own diet, Donna tried to explain that even though she was "heavier," she did not necessarily eat tons of sweet foods. This comment speaks to the stigma around heavier bodies, with others readily making assumptions about their lifestyles, habits, tastes, and lack of willpower. Donna also noted that she and her husband had tried different dieting strategies, like writing down their food and menu together. Her relationship with her weight and dieting had a longer history than this first intervention session.

For Donna, there was a lot at stake in maintaining a healthy diet for her and her baby. Both Diana and Donna drew explicit connections between Donna's diet and the baby's development. As Diana noted, "What you experience the baby also experiences." In Diana's framing, there is no separation of the maternal-fetal subjectivity; the fetus is described as being able to "experience" spikes in sugar metabolism. Diana approached the fetal-maternal relationship from a glycemic control perspective, and Donna referenced the fetal-maternal relationship in relation to risk and responsibility. Donna did not want to risk conception and pregnancy and would do anything to try to ensure the safety of her fetus. The complexities around weight, diets, sugar control, and pregnancy were magnified in Donna's experience. Both Donna and Diana came to similar conclusions about the effects that eating and food have on the health of the fetus. Donna's need and desire to control her weight—by any means necessary— were directly related to a desire to manage risks to the developing fetus.

### Ashley: Being Good and the Weight of Labor

For other participants, the intervention interacts with existing approaches to diet, food, and behavior. For instance, in Ashley's first visit with Diana, Ashley commented that she did not "eat badly."[40] Ashley knew what foods are "healthy" and which foods are "bad," so she felt like she didn't have much to work on during the intervention sessions. She continued explaining at her first visit with Diana: "I have a problem with chocolate and crisps, which I'm trying to compensate [for] with more exercise, but my biggest challenge is that I work from home." Ashley is a computer programmer, lives with one other flatmate, and primarily works from home. Ashley is a woman in her mid-thirties, identifies as white British, and was pregnant for the first time. In the first visit Diana listened and tried to assess Ashley's diet or regular eating habits. From Ashley's own experiences with other programs, she identified chocolate and crisps as "bad" foods, and it was also from other experiences of avoiding or restricting certain foods in other diets that she reflected on how these "bad" foods were problems for her.

During Ashley's second visit she shared her physical activity goals and commented that it was hard to walk after forty minutes. They also went over her goals of not eating out and limiting dessert, which again seemed similar to dieting goals. Ashley explained how she had eaten since the last visit with Diana.

ASHLEY: I had chocolate today, I have to be honest. My flatmate is a bastard. She brings in rubbish and I eat it.

DIANA: What can you swap for these things? Let's look at lesson 2.

ASHLEY: I drink water, I don't do rice, I don't do white bread, I do egg noodles, I do couscous, brown pasta. Breakfast I do porridge. . . . To be honest, it is just chocolate, just the odd chocolate. I have fruit and yogurt. I'm not into biscuits, cakes, or dried fruit. I am getting into frozen yogurt.

DIANA:  So for this week, on average how many [chocolate bars] will you have per week?

ASHLEY:  Three times per week. I had one bad day where I progressively had the whole package of chocolate. It was so good. The baby kicked the whole time!

DIANA:  Oh because of the caffeine, obviously baby kicking. Here are the measurements of caffeine. If you buy it you'll eat it. . . . The reality is that you are home a lot—it's good to stop you from buying it and bringing it in the house. Talk to your flatmate.

During this second session, Ashley seemed to talk more about what she did and did not eat, which did not allow Diana to get through the lesson. Ashley's past experiences with restricting or avoiding certain foods are apparent in her list of what foods she "did." Her moments of honesty reflect an internalization of "good and bad" behaviors or "good and bad" choices related to certain foods. Discerning between foods that are "good/bad" is a practice that Ashley was familiar with before enrolling the intervention.

Ashley's moralization of good and bad behaviors or foods existed before she started participating in the trial, which reflects her experience with dieting in the active and restrictive sense of the word. Her moralization of food is related to literature that discusses a neoliberal regime that frames people's eating decisions as "good or bad."[41] Annemarie Mol identifies this kind of framing as the "control vs. pleasure paradox."[42] Across both domains of good/bad and control/pleasure, I find that participants associate being in control with being good, and that indulging in the pleasure of eating is bad. For instance, Ashley's "problem with chocolate" is an issue of controlling or restricting how much chocolate she eats. Ashley claimed that she had a "bad" day when she ate an excessive amount of chocolate. At a different visit Diana asked Ashley, "How is the chocolate?" Ashley replied, "I hadn't had chocolate all week, until yesterday, and for me that's a bloody miracle! [. . .] I still feel like allowing myself that one blowout per week on chocolate, but this is not a diet it is a lifestyle." Her desire, motivation, and feelings about chocolate were not necessarily addressed in the intervention. Instead, Diana could only make suggestions about "healthy swaps" to replace chocolate. Diana emphasized different alternatives and also incorporating small snacks throughout the day so that Ashley would not be tempted to overeat in one sitting. The tension around chocolate as "bad" but also "so good" reflects the contradictions within the food as a substance that is nutritious, the controlling or restricting of food as morally "good," and the enjoyment or pleasure of food as "bad."

The other issue underlying the pleasure and control of eating and being good or bad relates to fetal risk. In the context of pregnancy Ashley's relationship to food and her desires were framed as risky for fetal development. In reference to the previous conversation about eating "loads of chocolate" in one sitting, Ashley commented that the baby was kicking in her stomach while she was eating the chocolate. Diana made the connection that the baby was kicking because Ashley was eating chocolate, which has sugar and caffeine. The exchange also made an explicit connection between Ashley's eating chocolate and the "baby's" response. Similar to the previous invocations of the maternal-fetal relationship, Ashley's behaviors were directly associated with her fetus's well-being. Ashley made the connection in an offhand manner, but it was Diana who brought attention to the immediate effects that Ashley's food choices and desires could have on the health of her baby. In the context of this trial, Ashley's "chocolate problem" was related to the potential risks of GDM and high BMI. A reductive interpretation of epigenetic logics suggests that eating too much chocolate can impact fetal development and future generations.

When Ashley returned to complete her last intervention session with Diana, she was twenty-six weeks and five days pregnant. Her glucose tolerance test (GTT) for the StandUp trial was scheduled in the next two weeks. In the last session her relationship with chocolate emerged again, and this time there were more connections made with how she was feeling.

DIANA: So how are things going?

ASHLEY: I can't sleep, the heat will kill me, but I'm still swimming. . . . I focus on swimming more and walking on another day and yoga. I wasn't so good with diet last week, bad day, cake, biscuits, and chocolate. I was emotionally down so I had comfort food. I was cramping, couldn't feel baby, and I was getting so worried. . . . I know it's about pulling myself out of this funk, but my partner is out of the country.

In response to Ashley, Diana reminded her of how well she had done in the intervention and that it was okay to have an "off week." Ashley experienced a wide range of physical and emotional changes that affected her mood, energy levels, and eating behaviors, as well as personal challenges. The fact that she could not "feel baby" instigated anxiety, and she did not have emotional support at that time. She was aware that emotional-physiological entanglements motivated her eating; however, her awareness also came with a harsh judgement of her behavior. Diana responded in a kind and empathetic manner, which is not technically described in the health trainer manual.

In the preceding exchange there is a nuanced and relational illustration of fetal-maternal interaction. In previous visits Ashley noted how her eating stimulated a fetal response of "kicking." In this instance the lack or absence of fetal movement stimulated a response in Ashley. She was worried and feeling emotional, which affected her eating behavior. What emerges from this interaction is that the fetal-maternal relationship is not necessarily unidirectional. It is possible that the behavior or movements of a fetus can also affect a pregnant person's feelings, behaviors, and eating. Similarly, but in a biological or material context, nascent research in epigenetics is finding that fetal particulate matter that seeps across the placental tissue can also activate or epigenetically modify maternal health outcomes like heart disease or preeclampsia in the pregnant/maternal/surrogate body during and after pregnancy.[43]

Nevertheless, no matter how complex the variables are around nutritional epigenetics and weight gain during pregnancy, in the end, all of this information was distilled into one goal for Ashley:

DIANA: Now you have awareness. In terms of goals, what would you like to work on in the future?

ASHLEY: My key goal is not to gain more weight, keep exercise going, and keep managing my workload.

After Ashley made the connection between her pregnancy, her feelings, and her eating, she concluded that her main goal was "not to gain more weight." What emerges as a dominant concern from these complex feelings is her weight. The burden of laboring through these feelings, anxieties, and responsibilities during pregnancy is not accounted for in the scientific framing of the intergenerational risk of chronic disease. Her body, her weight, her behaviors, and the food she eats all converge to represent the future risk and health of her developing fetus. Whether the language is framed around glycemic control or weight control, the same target comes into focus: pregnant bodies and behaviors are responsible for future risk. In turn, women take on the individual responsibility that is mistakenly aimed solely at their bodies. Nevertheless, their desires matter: they want to lose weight, they want to "be good" and control their diets.

## POLITICS OF POSTGENOMIC REPRODUCTION: THE DURABILITY OF CONTROL AND RESPONSIBILITY IN THE TWENTY-FIRST CENTURY

The themes of control and responsibility that emerge through epigenetic models of the maternal body help characterize contemporary politics of postgenomic reproduction. For instance, Sarah Richardson draws from literature in epigenetics and DOHaD to highlight a key contradiction in epigenetic models that focus solely on the maternal body: "[Epigenetic] changes manifest at the level of the intergenerational lineage rather than the individual female. The significance is that DOHaD research advances a shifting and mixed message regarding maternal agency and responsibility: it exhorts mothers to make lifestyle changes in the service of their genetic lineage, while maintaining that these changes are unlikely to bring them or their offspring any benefit. At the same time, it produces a model of the maternal body that suggests that maternal experiences, exposures, and behaviors may have very significant, amplified consequences for her offspring, her descendants, and society at large."[44] This excerpt throws into relief a key contradiction related to responsibility and control: epigenetic changes can manifest on the intergenerational scale, not the individual scale. Epigenetic modifications occur across time and space and are indeterminate and unpredictable.[45] Yet the message implicit to interventions during pregnancy is that individual women have the control and responsibility to change fetal development if they manage and discipline their behaviors. Postgenomic reproductive politics increases the stakes of individual present-day behaviors to intergenerational health outcomes: eating a donut *today* or gaining too much weight *this week* is directly connected to the future health of your children and their children. This imagined future is caked in anticipatory fear and anxiety that is embodied in pregnant people's experiences today.

However, Richardson emphasizes that individual women do not have "conscious control" over the unpredictable changes that can manifest in utero.[46] Richardson and colleagues caution scientists against "blaming the mother" or framing pregnant bodies as responsible and in control of epigenetic changes.[47] In an individualistic, control-oriented society, it is terrifying to admit that we do not have control over environmental exposures and epigenetic modifications during gestation. What you eat may or may not impact fetal development, and fetal development may already carry latent imprints carried on from past generations.

The unpredictable and indeterminate aspects of epigenetic modifications are often ignored in the application and examination of studies that target pregnant bodies. The potential is there, but the manifestations remain *unpredictable and indeterminate*. To circumvent or elide the unpredictability inherent to epigenetics, pregnancy trials focus on the behaviors of individual pregnant bodies in the present to prevent future health problems. The framings of health risks in the trials that I observed do not recognize the indeterminacy of modifications that can be stimulated by intersecting scales of the environment. For instance, in chapter 1 I discussed how the famine studies, which are foundational to DOHaD models, found patterns linking health outcomes to grandparents' nutritional exposures on the *paternal* side. The manifestations can skip a generation, or be carried on, across maternal or paternal lineage; therefore, intervening during pregnancy may or may not affect the immediate generation.[48]

The politics of postgenomic reproduction reveal that pregnant people are placed in an impossible position: they are deemed responsible and in control of their behaviors, as if their bodies and behaviors have significant and irreversible impacts on epigenetic modifications in utero. Such foreclosed interpretations of epigenetics and the maternal environment, as only individual pregnant bodies and behaviors, selectively ignore the inherent theories of indeterminacy in gene-environment interaction and regulation. Consistent with these foreclosed interpretations of epigenetics, contemporary pregnancy trials frame cis-gendered pregnant bodies as discrete individualistic forms of the environment, further described in the next chapter, that solely control the nature, nurture, and nutrition of fetal and infant development.

This is not to say that what you eat during pregnancy is inconsequential. Nutritional guidelines are important *to a certain extent*. There is no scientific consensus on what pregnant people should eat or how much they should weigh.[49] Different cultures apply and interpret prenatal nutrition guidelines in distinct ways. The categorization of "risky" foods during pregnancy also varies across national contexts.[50] Regardless of the variation and lack of scientific consensus, I do not promote the entire dismantling of maternal nutrition guidelines. The point is that current scientific understandings cover a small fraction of the story related to food and eating during pregnancy, and that focusing solely on weight through methods of control, responsibilization, and restriction during pregnancy does not significantly improve pregnancy or infant health outcomes in the present.

You could have the most "perfect" and "healthy" diet, determined primarily by your social location, class, and education, and still your children might develop metabolic syndromes. Adverse exposure during utero or any other critical period does not deterministically result in a sick future or a healthy future. I emphasize that there is *indeterminacy* in how toxic racist environments shape us. Whether you ate "healthy" or your mother, father, grandmother, or grandfather ate "healthy" does not predetermine your health outcome in the present, it may give us more information, but how that information is interpreted and used is directly connected to the epistemic environments of racism, late capitalism, and late liberalism.

Another aspect that is often ignored is that epigenetic models and justifications for lifestyle interventions during pregnancy are not applied to a blank slate. The focus on controlling and managing pregnant bodies in efforts to protect future citizens—an enduring aspect of twentieth-century biopolitical and reproductive politics—is embedded in structures of medicine and research and internalized at the individual level. I found that it is not just the scientific community that frames pregnant bodies as being in control and responsible. Rather, the pregnant narratives explored here also reflect that individual women assume responsibility for their diets and for how their diets affect the developing fetus.[51] Participants take on responsibility *as if* they were in total control of their children's health and biological development.

Large, state-endorsed hygiene campaigns indicative of nineteenth- and early twentieth-century strategies are no longer necessary to discipline pregnant people into behaving like good healthy mothers. In a postgenomic, fat-phobic, late liberal and capitalist era, pregnant participants in the United States and United Kingdom have already learned that nonnormative bodies are ugly, risky, and unhealthy; if they do not seek out interventions, commercial or medical, they are irresponsible parents and will be held accountable by society and the state. Furthermore, the marketization of reproduction promotes culturally hegemonic narratives around "good" parents, characterized by and through responsible consumptive behaviors. In our current social and economic climate, being poor and a person of color diminishes the capacity to be a "good" parent; what you eat and how much you weigh during pregnancy further exacerbate the stratification of reproduction.[52]

Those who decide to enroll in nutritional interventions during pregnancy represent a part of the population that looks to lifestyle pregnancy trials for support to bear the weight of social and medical stigma associated with being an overweight, diverse, pregnant person in a fat-phobic, racist society. In exchange for such support, participants subject themselves to extra surveillance and management in lifestyle pregnancy trials. For participants in these pregnancy trials, whether epigenetics or any other theory, the assumption is the same: "mothers" are solely responsible for the health of their children.[53]

·    ·    ·    ·    ·

This chapter is one of the first ethnographic explorations of pregnant participants' experiences in ongoing lifestyle interventions. It has revealed the fears and anxieties that pregnant participants have about managing their weight during pregnancy. The themes of control and responsibility emerged as significant in shaping not only the design and justification of lifestyle interventions during pregnancy, but also in individual pregnant women's own motivations to enroll in these trials. The pregnancy narratives from contemporary lifestyle interventions also reflect how notions of control and responsibility are enveloped in epigenetic models.

I included a variety of perspectives from participants and staff members in the trial to show how pervasive the experience of weight control is for pregnant and nonpregnant women. In addition, I examined how the content of the UK intervention was similar to commercial diet programs, yet the trial claimed that the intervention was "not a diet for weight loss." Regardless, the trial functioned like a diet program for the participants, one that also framed certain foods as "good or bad," "healthy and unhealthy." Overall, the chapter shows how contemporary lifestyle interventions are challenging to implement because each participant has a deep emotional and physical interaction with what they eat and how much they weigh, all of which is compounded by economic inequalities like housing and employment that significantly impact individual behaviors like eating and buying food.

Shaina's, Mary's, Donna's, and Ashley's encounters in the intervention sessions reflect the broad and complex spectrum of experiences that I observed from ethnically diverse participants enrolled in the UK and US pregnancy trials. For the participants, being "good" and "healthy" meant controlling, restricting, and managing their bodies, behaviors, and desires. One of the motivating factors for enrolling in the trial was to manage their weight through a "diet" program, whether it was in the intervention group of the trial or a for-profit commercial program. Diana, the health trainer, understood these complex desires, and she translated both the goals of the trial—"this is not a diet for weight loss"— and the participants' own needs. Furthermore, focusing on weight as a main indicator of health elides the potentially "unhealthy" ways in which people try to attain "healthy" bodies. Donna's experience with managing her weight by any means necessary is reflective of an implied message: losing weight is more important than *how* you lose it. Nevertheless, participating in these prenatal intervention programs helped the participants meet the social and medical expectations of being, good, responsible, healthy mothers. It does not matter that nutritional interventions during pregnancies do not necessarily impact pregnancy outcomes or infant health outcomes; for these participants, it is important to play the role of trying to lose weight. The belief that losing weight is synonymous with improving health is so ingrained in Western medical systems that no alternative solutions are even imaginable.

A key point that emerges in the participants' narratives is how diets and nutritional or "healthy" ways of eating are political and racialized. For Shaina and Mary, both first-generation, Black British women, the intervention represented an indoctrination into "healthy eating," one that is entrenched in colonial and racialized regimes. In this way, the nutritional intervention was also not a neutral scientific tool. How glycemic control is mobilized and coupled with Mediterranean diets reflects another side of prenatal nutritional interventions, one that exposes the racial politics inherent to food and science. If we are what we eat, or what our grandparents ate, our bodies are materially enmeshed with the stratified politics of food. Discipling people into not speaking their language, or sending them to reform schools, is akin to disciplining people into eating the "right" (white-European), healthy way. Reproducing whiteness is not just biological, but also social and cultural. As such, reproduction is and has been a target for racist and eugenic interventions precisely because it is at once biological and social (re)production.[54] Familiar dynamics from twentieth-century reproductive politics are enlivened in new ways through twenty-first-century pregnancy trials.

The on-the-ground realities for pregnant participants in the intervention remind us that scientific paradigms may change, and new theories can emerge, but their day-to-day experiences are shaped by epistemic environments that are characterized by hegemonic, fat-phobic, racialized body norms that make them individually responsible. The postgenomic shift did not change the underlying pressures that pregnant women feel to change their bodies and behaviors for the sake of their unborn children, even though epigenetic science states clearly that pregnant women cannot control epigenetic modifications in utero. In this way, the focus on individual control and responsibility remains significant in prenatal lifestyle trials and indexes the contradictions inherent in the contemporary politics of postgenomic reproduction.

# Part III

# 5  Environmental Animations

You are twenty weeks pregnant, and you have your first intervention appointment for the clinical trial.

You go to the same hospital for all your appointments, but this time instead of taking the elevator up to the eighth floor, you go to the fourth floor. This floor is like a maze, with lots of rooms. It's gray and cold, and feels like a laboratory.

You sit and wait in the administration center, then see a person walking toward you with a big smile. She is the interventionist for the StandUp trial. She leads you to a room with a bed, some medical equipment, a computer, and a chair.

After some warm introductions, she explains that the purpose of the intervention is to help you make healthy dietary changes. Then she tells you a story about sugar metabolism: "Let's say you eat a donut in the morning. You chew, swallow, and the fried dough and sugar moves down your throat and into your gut. Once it gets into your stomach, it is quickly absorbed into the blood because it is easy to break down. Simple sugars or carbohydrates are quickly released into the body. Once in the bloodstream the sugar stimulates the pancreas and liver. These spikes in blood sugar are not healthy for you or the babe. Whatever you experience, the baby also experiences."

She shows you a graph with a line that sharply spikes up from low to high.

You're startled, because you remember picking up a puff pastry on your walk to work yesterday from the local bakery. The bakery is next to the oil refinery. You start to worry about the pastry, but not the years and generations you and your family have lived and worked next to the refinery.

.     .     .     .     .

During their first intervention session in the UK trial, the participants receive an explanation of sugar metabolism that emphasizes food and the maternal environment as key factors for fetal development. Participants would learn about blood sugar spikes and how they impact the "mum and the babe." The interventionist would use the terms *mum* and *babe*, but the PI of the trial would say "maternal metabolic environment and fetal development." This approach draws from nutritional and environmental epigenetics, which claims that food can act as an environmental stimulus to influence fetal programming, as some epigeneticists call it, or fetal development, as described in chapter 1.[1] Pregnant participants did not use the term *maternal environments*; their conceptualization of the environment differed from that of the scientists. And no one in the trials ever mentioned stress, pesticides, or lead exposure. To illustrate how mothers outside the context of the prenatal trials understand epigenetics and obesity during pregnancy, I explore an encounter I had with a mother of two after a research workshop.

In March 2016, I gave a presentation at an international workshop in Europe. In this talk I discussed the significance of nutrition in epigenetic modification, and I also talked about the importance of context and the entanglement of environments. I introduced the hotly debated topic of transgenerational and intergenerational inheritance and how these new theories of inheritance focus on the role of prenatal nutrition in health across the life course. I also explained that epigenetic modifications are unpredictable and influenced by multiple entangled environmental factors.

After my presentation a white woman in her mid-forties from France came up to me to talk about my research. She prefaced her question with a story. The woman, whom I call Annie, had a very traumatic first pregnancy. When I asked her why it was traumatic, she told me that her doctors kept telling her that she was gaining too much weight, and she was very worried. To try and limit her weight gain, she walked every day and watched what she ate. Annie avoided desserts, monitored her weight, and avoided eating too much, but the number on the scale kept going up. Annie felt very helpless and out of control. Her entire pregnancy was marked by fear of gaining too much weight.

At the end of her pregnancy Annie gave birth to a healthy eight-pound baby. Fast-forward seventeen years: Annie tells me that her first born is struggling with her weight. Annie's daughter cannot stop eating sugar and does not have the taste or desire for anything else. "She just eats sugar," Annie said. Then she finally asked me the question: "So did I do this to her? Did I make her want sugar because when I was pregnant I was so worried about eating sugar and gaining weight during pregnancy?"

I stood there and looked at Annie, who was looking back at me, awaiting my response. My initial reaction was to say "no," but I knew that would not satisfy her. She wanted me to confirm her suspicion in order to come to terms with the guilt and responsibility she felt about her daughter's current eating habits and desires. I wanted to tell her that her feeling of responsibility and guilt for having "done this to her daughter" was based on a long historical and political framework that has influenced Annie's belief that she is solely responsible for her daughter's eating habits. What I did explain was that the risk and responsibility with regard to developing children and populations relates complexly to social, political, and environmental exposure. Moreover, I explained that there is no possible way that she could totally control or be solely individually responsible for the health of her daughter, because her body was also vulnerable to environmental stimuli while pregnant. As the last chapter described, epigenetic modifications and how they are inherited across generations lie predominantly outside individual control. Finally, I responded that epigenetic modifications can be unpredictable and can be triggered by environmental factors that are beyond one's own individual body and behavior.

Mothers like Annie and many scientists underestimate how much people's lives are shaped by scientific speculation. In conjecturing on how exactly pregnant behaviors may epigenetically impact fetal development, the importance and instability of contexts and scales of the environment are selectively ignored. Although epigenetic science claims that women who are obese during pregnancy have a higher risk of having children who will be obese and diabetic, epigenetics also claims that the risk is contingent and unpredictable. In scientific discourse, the indeterminate aspects of epigenetic modifications are framed as a problem to solve. Yet this indeterminacy is also crucial to epigenetic potential. Embracing the indeterminacy can shift attention away from overdetermined assessments of an individual's predisposition to disease, assumed to be caused by individual behaviors. Despite the role of environmental factors that influence health outcomes and the importance of *paternal* biology and behaviors, the interpretation of epigenetics by Annie as well as scientists in implementing prenatal interventions focuses entirely on pregnant bodies and behaviors.

In this chapter I argue that there are many different kinds of environmental exposures that can affect fetal development, but only certain kinds of environments are animated and targeted in the prenatal trials I examine in the United States and United Kingdom. Assumptions about where the maternal environment begins and ends are precisely what is at stake when comparing control and experimental groups in these pregnancy trials.

This chapter is oriented around these questions: What counts as the maternal environment in contemporary pregnancy trials? How do different experts frame environmental targets for intervention? And what implications result from the privileging of certain environmental factors for data collection? In examining these questions, the chapter illustrates a valence of epigenetic foreclosure indexed by how the maternal environment is imagined in postgenomic pregnancy trials. Before exploring the ethnographic material, I review some definitions of the "environment" in epigenetic science and its impact on making "maternal environments" the main targets for intervention.

## THE "ENVIRONMENTS" IN EPIGENETICS

As I introduced in chapter 1, the growing scientific and public attention around epigenetics has brought forth a renewed focus on the environment as a significant factor in genetic development and inheritance.[2] I say renewed because the role of the environment in genetic development has had a capricious history that can be traced in and out of evolutionary science.[3] I use the term *the environments* to emphasize how epigenetic environments are framed as one factor and many factors all at once. Epigenetic theories underscore the multiple and intersecting scales of environments, while application of epigenetics in pregnancy trials focuses on discrete individual factors. Epigenetic foreclosure illustrates how multiple coexisting environments become selectively translated and examined as one environmental factor.

For instance, scholarship on epigenetics shows how concepts of the environment are animated through notions of exposure, scale, space-time, and molecularization.[4] In current epigenetic literature, the environment encompasses that which is both inside and outside the body.[5] Epigenetics emphasizes that the environment occurs on different *scales*, from the cellular level all the way to the atmospheric level, including cells, uteruses, the microbiome, and the natural built landscape.[6] In this multiple and iterative approach to the environment, food can also act as an environmental factor. Hannah Landecker explains that the reconceptualization of food as an environmental factor indexes a fundamental shift in how epigenetics has changed our understandings of the environment, but not necessarily of food.[7] Landecker argues that food as exposure has existed in one way or another throughout history; epigenetics is just one particular moment in our social and cultural understanding that frames food as a type of environmental exposure in a slightly different way. The key difference now is that nutrition is framed as an environmental factor that can stimulate epigenetic modifications into the future.[8]

In my analysis, what counts as the environment is also directly related to the biomedical frameworks applied in measuring and tracing environmental factors to changes in genetic expression. Different approaches to epigenetics will focus on distinct areas of the environment. For a neuroscientist who studies epigenetics and brain plasticity, the environment can include daily experiences.[9] For molecular biologists, environmental stimuli can refer to biochemical material like methyl groups that attach to DNA sequences.[10] Moreover, for a scientist working at the National Institute of Environmental Health Sciences, environmental agents are defined as "mold, pesticides, air pollution and some foods and medications."[11]

How we study the environment is a political project with significant implications for maternal health. The various ways that environmental targets are rendered visible and intervenable in the application of epigenetics also create certain forms of responsibility and epigenetic potentials and not others. Within the prenatal trials that I study, one form of epigenetic foreclosure occurs when reductive and selective animations of environmental factors are used to justify prenatal interventions.

## RESPONSIBLE ENVIRONMENTS

The scholarship in epigenetic science frames the environments as multiple, porous, scalar, and spatiotemporal, but in practice this proves difficult to methodologically study in randomized clinical trials. In the trials that I examine, pregnant people were not explicitly told that their diet was an environmental factor. Still, the scientists drew from nutritional and environmental epigenetics to justify the significance of prenatal interventions precisely because food and pregnant bodies were both assumed to be environments of interest to the study.

Pregnant participants in the trials ate certain foods *and also* lived within environments of stress, pollution, violence, and racism, social phenomena that are hard to control for in an randomized clinical trial (RCT). Michelle Murphy's concept of distributed reproduction emphasizes that racism and settler colonialisms shape reproduction and the unequal exposure to toxic environments.[12] Pregnant bodies are not immune to or protected from the ubiquitous pollutants that linger in our bodies and ecological systems across generations. As Lappé and colleagues explain, the significance of environmental activism and reproductive justice was made meaningful in new ways after 2000, when epigenetic mechanisms and DOHaD theories joined forces to corroborate the impacts of environmental exposure to health outcomes across the life course and intergenerational transmission of epigenetic imprinting.[13]

Although maternal environments are also multiple and entangled with broader scales of the environment beyond pregnant bodies, pregnancy trials employ individualistic framings of the maternal environment. The main environmental factors that are studied in pregnancy trials are pregnant bodies and behaviors. The hyperfocus on maternal environments as critical periods for fetal development also justifies targeting pregnant populations for interventions because they are assumed to be the only environments responsible for adverse health outcomes in future generations.[14]

Despite the complex role of environmental factors influencing health outcomes, scientific discourses often collapse the whole environment to the maternal environment and target the maternal environment as the first source of environmental exposure for developing fetuses.[15] By examining reproduction and epigenetics in South Africa, Manderson finds that the increased focus on women as "foetal containers" reinforces the idea that women are "vehicles of poor intergenerational health."[16] Similarly, the scientists who draw from epigenetic science to test nutritional interventions on obese pregnant women frame pregnancy as a "window into the health of future generations."

My intervention in the social and theoretical examination of epigenetics focuses on using ethnographic data to illustrate the epigenetic foreclosure of the environment in epigenetics and its impact on reproduction. I spotlight the ways in which epigenetic knowledge and practice can both reinscribe older prescriptive models of genetics and race *and* provide an entry point for challenging those same models.[17] Thus, I study epigenetics as a political and scientific theory in relation to race and gender. My approach to epigenetics foregrounds what happens when multiple and intersecting scales or domains of the environment are in play.

Here, I explore the clinical experience of the environment: specifically, how different scientific approaches and interpretations of epigenetics turn particular domains of the environment into targets for intervention. I show how the application of different biomedical frameworks creates particular environmental targets and shapes the design and implementation of prenatal interventions. In what follows, I focus on three domains: pregnant bodies as environment, home as environment, and everyday experiences as environment. The flexible definitions of the environment are significant in the clinical translation of epigenetics because what counts as the environment influences the sites and types of interventions tested in evidence-based medicine and the type of data collected and deemed valuable for scientific discovery. Moreover, the determination of what constitutes a "risky" environment (requiring intervention) generates material and bodily consequences for pregnant people (such as invasive medical surveillance). The way in which the environment is rendered visible and intervenable impacts distribution of reproductive responsibility and accountability.

## PREGNANT BODIES AS ENVIRONMENTS

One of the first ways that environments came into view in the studies I observed could be seen in the animal models used as a basis for the clinical trials. The prenatal trial in the United Kingdom was based on studies of pregnant rats as environment and treated pregnant bodies themselves as environments. The principal investigator (PI) at the UK trial, a physiologist by training, explained at a conference:

> We do a lot of our work in animal models. We go backwards and forwards between animal models and the clinic. And these models are incredibly important to us. For our animal models we give rats and mice absolutely delicious things to eat, and then they get fat and then we make them pregnant, and this is a good model of obstetric obesity. And we've been looking at the *children* when they grow up, or the offspring of these rats and mice, and it's extraordinary. They have very high blood pressure, they become fatter, and they have abnormal glucose control. Therefore, we believe that the fetal development is very susceptible to the *maternal environment* and that it predisposes children to disease as they are exposed to these metabolic conditions in utero (my emphasis).

The PI cites her published research to argue that animal models provide evidence for the adverse effects of maternal obesity on *human* offspring.[18] The use of animal models to understand environmental epigenetics in humans is a common practice.[19] For instance, Michael Meaney and Moshe Szyf, geneticists at McGill University, use animal models to make the case that "maternal care" can influence fetal programming.[20] By showing how the PI uses animal models in the UK trial, I emphasize how her methods and expertise shape her understanding of environmental targets for prenatal interventions in humans. It is through the animal models that certain environmental targets are constructed as important in the design and implementation of the prenatal interventions. As a physiologist interested in metabolic disorders, the PI focused on how the maternal metabolic environments of the rats could affect the development of fetal mice.

In the UK rat models, maternal and metabolic environments are prioritized, while other aspects of the environment disappear. What about the temperature of the laboratory, or how scientists handle and force-feed the rats? Those aspects are also a part of the maternal rat environment and the rat fetus's environment. The underlying assumptions involved in translating animal models to human interventions, or assuming the standardization of animal environments in laboratory settings, is a classic concern in science and technology studies (STS).[21] For instance, Muller and Kenney point out that the animal model mentioned by Meaney and Szyf focused entirely on motherly behavior (characterized as licking and grooming), which "black-box[es]" other aspects of the environment like the "cage, food, the other rat pups."[22]

By ethnographically examining the clinical translation of the rat models to human interventions, I find a similar set of insights. For instance, like the rat models, other aspects of the pregnant participants' environment, such as housing, toxic exposure, and stress, did not count as key environmental targets in the clinical trial implementation. In the prenatal intervention, the aim was to change the food that was entering the pregnant body, which affected the maternal metabolic state and the intrauterine environment. By standardizing and controlling other scales or domains of the environment in the lab, the pregnant body emerges as the key environmental target for intervention. In this way, the animal models show us what selected parts of the environment emerge as significant and how those selected factors are then correspondingly applied or assumed in clinical experimentation on pregnant people.

The point I underscore is that while Meaney, Szyf, and the UK PI all use animal models to focus primarily on the maternal environment, a key difference across the studies is what emerges as the significant aspect of the maternal environment to target. Meaney and Szyf are geneticists, and they selectively focused on the licking and grooming as key environmental factors that influence genetic expression in the pups. The UK PI is a physiologist by training, and her focus is on the maternal metabolic state of the rat, which is why the rats are force-fed to replicate an "obesogenic" maternal environment. The environmental targets are different in each study because they are approaching, interpreting, and translating the animal models in distinct ways based on their distinct expertise.

The different ways in which environments are conceived directly affect the kind of interventions that are designed. In the Meaney and Szyf study, they claim that intervening in "motherly behaviors" can impact (and potentially even reverse) epigenetic modifications in genetic expression.[23] In the UK animal models the environment that is targeted for intervention is the maternal metabolic environment, which influenced the PI's choice when designing a nutritional intervention for pregnant women.

In her own words, the UK PI explains that through her animal models she has found "that the *maternal metabolic state* plays a very important role in the future risk of disease in the developing child [. . .] . The suggestion, which is not unusual, is that the *intrauterine environment* has a prolonged effect on the health of the child" (my emphasis). The PI's understanding of environment incorporates a temporal as well as a spatial component. The "future risk of disease" and the "prolonged effect on the health of the child" reflect a temporal understanding of how environmental modification can be carried across time and space. Two key assumptions that undergird the PI's notion of epigenetic environments: one is that scales of the environments can be controlled and standardized through experimentation, and the other is that the pregnant body can be studied as a temporal environment that extends into the future.

In addition, by zooming in on the metonymic relationship between parts of the maternal body and the whole maternal body, I characterize the reductionism, a process that matters for understanding how exactly pregnant bodies become the single environmental targets for intervention. For instance, the "maternal metabolic state," including glucose metabolism, excess fat, cholesterol levels, and blood pressure, is also part of the "obese" pregnant body or maternal environment. Likewise, the use of the term *intrauterine environment* is also a part of the larger maternal environment. Therefore, what constitutes the maternal environment is differently framed in the process of focusing on the metabolic and intrauterine domains, as opposed to "motherly care" or other gendered ideas of the maternal environment. Furthermore, what counts as the "maternal environment" also shapes the various interventions tested in the clinical translation of epigenetics. In the UK trial this manifested as a prenatal intervention that focused on glycemic control, which attempted to impact the maternal metabolic state and intrauterine environments.

The relationship between parts of the maternal environment and the whole maternal environment is selectively rendered visible through the scientist's expertise and use of animal models. This particular approach emphasizes pregnant bodies as environments for fetal development, rather than bodies situated within multiple scales of the environment. Barbara Duden (1993) and Monica Casper's (1998) work offers ways of understanding how the "fetus" becomes the main focus while the pregnant person disappears from view by being reduced to an environment or space for fetal development.[24] Feminist scholars note that anti-abortion, pro-life politics impact the way the fetus is foregrounded in maternal health, which promotes fetal personhood as a legally legitimate subject at the expense of maternal rights. Overall, the selective prioritization of the maternal metabolic state, intrauterine environment, or fetal environments obscures women's (and rats') lived experiences—experiences that are not part of the environment that matters in these particular applications of epigenetic science.

## HOME AS ENVIRONMENT

The PI for the US trial is a health psychologist by training, and she offers a different set of insights on what counts as an environment in her studies. Whereas the UK trial focused on the metabolic state or intrauterine environment, the US trial focused primarily on the idea of the home environment. In addition, the US PI did not use animal models, instead applying her expertise in nutrition and health counseling to design and implement nutritional interventions on pregnant women who were classified as overweight and obese. Twenty years ago, when the PI began her work on behavior, diets, weight loss, and pregnancy, the research was situated primarily in the field of health psychology. However, with the emergence of epigenetics, her work on maternal nutrition and behavior is spotlighted as a site to examine how nutrition during pregnancy changes genetic expression in fetal development *and* how maternal behavior during early development may also affect gene expression.

The main goal of the US intervention was to minimize weight gain. Following the recommendations published by the Institute of Medicine (IOM), pregnant participants in the intervention group were counseled to limit their weight gain to only a half a pound per week during their pregnancy. To meet the weight gain goal, the women in the experimental group had to follow a strict meal plan that included meal replacement beverages. In addition, each participant had to meet physical activity goals and meet with a nutritional counselor every other week throughout their pregnancy. The intervention was rigid by anyone's account; however, the participants were not forced to do anything, and compliance varied across participants. The strict meal plan aimed at limiting weight gain and calorie consumption. Exposure to food, or calories, was associated with weight gain, and different home environments indexed different kinds of food exposure.

In the protocol manual of the US trial is a section that outlines forms of measuring or assessing the "home environment." Participants complete a survey that asks them about the number of exercise equipment or pieces and the number of high- and low-fat foods in the home. The trial also assesses "food storage" at thirty-five to thirty-six weeks of gestation, at twenty-two to thirty weeks postdelivery, and then forty-eight to fifty weeks postdelivery. At these moments during and after pregnancy, participants are required to make a checklist of all the foods in their refrigerators and cabinets. The protocol manual also emphasizes that these surveys have been "used in diverse patient populations."

In an interview with the PI, she explained that "the intervention is an intensive environmental manipulation because we are *reducing exposures to food*, and so it is a pretty intensive approach to reduce intake of food—we tell them exactly what to eat, when to eat, and what not to eat, and give them something to use instead of eating" (my emphasis). The PI's use and framing of the "environment" echoes similar framings of food as exposure presented in Landecker's work; the main difference is the application of a healthy psychology approach. When I asked what exactly she meant by "reducing exposure to food," she responded by saying that, for instance, "having a meal replacement shake [instead of a sandwich], you're not having ham and cheese and bread, and mayonnaise, and mustard, and everything else that you could put on a sandwich, you just have this one thing, and you don't have all the excess food in the house." From the PI's perspective, she reduced pregnant women's exposure to food by literally reducing the availability of food in their homes. Food as exposure and exposure to food seem to be two sides of the same coin. Through epigenetics, if food acts as an environmental factor in the uterine environment, then reducing the exposure to food in the women's home environment is by extension another way to reduce the food that enters the uterine environment. As such, food as an environmental factor is meaningful on both the scales of the home environment *and* the uterine environment.

The US PI further characterizes the home environment as follows: "I think we can change their home environment and I think that is exciting because of the babies who are going to come and play and develop in that home environment. I think the women in the intervention will have a more healthy home, food, and exercise environment." For the PI, the home environment is a psychological factor in changing behavior, and within the epigenetic literature, it can also act as an environment that influences biological and social development. Following her description of the home environment, I asked her about how the trial specifically intervenes in women's behaviors and environments. "On a basic level, we are giving the women scales, measuring cups, pedometers. We give them some environmental cues. We are asking them to exercise, so that usually leads to more sneakers around the house, or gym clothes around the house and yoga tapes and exercise tapes. We can manipulate it here and there, but I think in this study it is more exercise cues, and some dietary cues, meal plans and shakes. That is manipulating their home environment. We also tell them to do a cabinet clean out; to take out all the junk from their home." In this excerpt the PI explains that tennis shoes, exercise tapes, or the types of food in the kitchen cabinets are all aspects of the home environment that can affect or "cue" behavior, which then affects what kind of exposure to food people have. The woman's behavior, along with what kinds of things she includes in her home environment, can by extension affect the food that the developing fetus is exposed to and the home environment in which the future child develops. In this way, the pregnant body is the intermediary environment between the fetal environment and the home environment.

In both trials, PIs render the environment in different ways, which make certain epigenetic environments subject to examination and intervention and not others. In the US trial the maternal environment was shaped by calories, and it included the home environment. In the UK trial the key design aspects of the nutritional intervention were informed by the animal models, which focused on the maternal metabolic environment. The pervading logic of the UK trial intervention focused on eating foods low on the glycemic index because those foods release sugar slower and directly affect the metabolic functions of the pregnant woman and fetus. A focus on calories and weight control does not necessarily address the substance or metabolic effects of foods, but from the US PI's perspective, lowering calories will indirectly affect sugar and fat consumption.

The different ways that experts approach, monitor, and categorize environmental targets shape the different kinds of interventions that are designed and implemented. For example, in the US trial the home environment and food as exposure were mobilized to justify the intervention of calorie and weight control. In the UK trial, the focus on the maternal metabolic environment resulted in an intervention of glycemic control, which does not directly restrict calories or weight but focuses on managing sugar metabolism. Regardless of whether scientists refer to home environments or intrauterine environments, it is the pregnant participant who is held to be responsible and accountable. Counting calories, controlling weight, and eating low glycemic foods all require the individual "mothers" to change their behavior. These environmental animations center the body and behaviors of pregnant people, which shapes the kind of data deemed valuable for collection.

The home environment rendered in the scientist's imaginary indexes a particular socioeconomic status. For instance, some home environments do not have the extra funds to buy exercise equipment or to fill the kitchen cabinets with calorie-controlled snacks or fresh produce. Assuming that the home or maternal environment is stable, safe, and controllable allows for the intervention to target individual behaviors rather than targeting interventions at the level of community and state policies. Framing the home or maternal environment through food, behavior, and choices enables the responsibility and accountability to be centered on individual bodies. If scientists were to imagine a different maternal environment, one that takes into consideration the multiple scales of intersecting environments at play, that would no doubt redistribute responsibility beyond the individual alone.

## EVERYDAY EXPERIENCES AS ENVIRONMENT

In the clinical trials I observed, monitoring weight, diet, and the food available in one's home environment was framed as an individual and behavioral choice. However, as Leith Mullings and Alaka Wali state in their work on reproduction in Harlem: "[W]hat appear to be personal risk factors and individual lifestyle choices are best understood in the context of a larger structure of constraints and social choices conditioned by race, class, and gender."[25] Their point is that environmental contexts are not only structured by gender, race, and power; they also shape material conditions of *exposure* for individuals and for pregnant bodies.[26] The "larger structures," which can also shape epigenetic modifications, are illustrated in greater detail through the experiences of the pregnant participants.

Based on my work as an interventionist and interviews with other staff at the SmartStart trial, the intervention visits were rarely about the food that the pregnant participants ate. That is, the intervention visit itself was entangled with the pregnant people's families, work schedules, transitions, unemployment, evictions, and childcare. The conversations during the nutritional interventions were less about calories or weight and more about managing life in general.

To illustrate the different experiences of the pregnant participants, I turn to an ethnographic case that I call Participant Iris. This ethnographic case illustrates similar issues that different participants in the SmartStart trial experienced, but to maintain patient anonymity I have modified identifiable details so that it cannot be traced back to one individual. Iris's case highlights some of the patterns I observed across different participants but does not intend to stand in for all of the women's experiences. Since I spent much of my time recruiting and delivering the intervention at the public clinics rather than at the private clinics, Iris's case is representative of the participants I observed who were from a lower economic class.

Iris is a woman of color who was deemed to be obese and volunteered to participate in a clinical trial. At the SmartStart trial, the majority of women I worked with and observed were women of color. Recall from chapter 3 that the trial aimed to recruit 50 percent Hispanic and 50 percent non-Hispanic women. In addition, the trial aimed to recruit an equal number of participants at different income ranges, however, this goal was difficult to implement due to unpredictable aspects of participant recruitment and retainment.

Iris was in her late twenties, had two kids under the age of four, and was pregnant with a third. She was active in her church and ate at the weekly potluck, which was one of the only cooked meals her family had each week. Money was tight because she and her husband were unemployed. Iris also suffered from depression and insomnia, and at twenty-four weeks' gestation she was diagnosed with gestational diabetes mellitus (GDM). Throughout the trial I observed symptoms of prenatal depression; I was trained to triage any serious cases to my supervisors, who were clinical psychologists. In addition, the trial identified a significant number of women with gestational diabetes. The higher rate of GDM among the participant population was attributable in part to the trial using a more conservative measure for diagnosing GDM compared to standard care. After Iris was diagnosed with gestational diabetes, she met with a dietician every other week and with a diabetes specialist. Altogether, she had about eight health appointments each month, including her participation in the clinical trial. Near the end of the pregnancy, she struggled with eviction, and eventually she and her family moved out of their apartment to live with some in-laws.

At the weekly staff meetings, I presented Iris's case and the significance of her mental, emotional, and physical living conditions, noting that she was suffering from housing insecurity. However, I was reminded that I had to focus on delivering the intervention as it is delivered to everyone in a standardized manner. This reflects what is measured and what is not, or rather what is considered an environment to target and what is not. For instance, food may be an environment, using the meal replacement shakes may reduce the participants' exposure to food, and the intervention may change their home environment, but the stability of their income and housing are not considered measurable environmental factors in the RCT.

Sometimes I felt conflicted asking about Iris's food journal, calorie goals, and steps she walked that week, because she had other environmental factors to worry about. She usually preferred to talk about the immediate and material challenges she faced every day, and I would listen. Despite her stressful days, Iris kept coming to meet with me every two weeks. In the end, she delivered what the trial classified as a "healthy baby," and she met the trial's weight gain goals. The study followed up with her six months after delivery to weigh her and her baby and take bio-samples for testing. The biomarkers they test indicate whether or not the nutritional intervention changed any genetic expression by reducing the baby's risk of developing obesity and diabetes. What remains unchanged are the participant's living conditions.

The trial was not designed to target the environmental factors that emerged as significant in Iris's everyday experiences. During our interven-tion sessions, Iris commented that fast food was often cheaper than fresh produce. When I asked her about her steps and how she could fit in a walk during the day, she told me that she did not feel safe walking around her neighborhood. In addition, meeting the physical activity requirement was difficult because Iris suffered from insomnia and depression, which made it difficult to feel motivated to exercise. Although it was laborious, Iris still kept attending the trial, and even though she did not completely comply, she made an effort.

As an interventionist in the trial, I felt that multiple understandings of the environment competed for my attention and care. On the one hand, I saw how Iris's environment and living conditions influenced her health and emerged as an active force in the intervention visits; on the other hand, I had a concept of the environment from the US trial that focused on food or calories and the home environment. Further, what counts as "the mater-nal environment" from the participants' perspective indexes an entirely different set of concerns that implicate a different set of interventions. More importantly, the view of the maternal environment from the preg-nant woman's experiences prompts us to reconsider the bodies that are held responsible and the environments that are targeted for intervention.

The ethnographic exploration of epigenetics illuminates how differ-ent frameworks shape what counts or does not count as an environment worth targeting for intervention. In different scientific settings, the target for intervention can be food, bodies, and behaviors. The hyperfocus on any one of these domains obscures the interconnectedness and multiplicity of epigenetic environments. The effort to use epigenetics in these trials as a way of predicting genetic expression in the future obscures the multitude of social, political, racial, and gendered environments that influence the

expression and possibility of health outcomes in the present. More specifically, focusing on changing what a woman eats during pregnancy in an
attempt to reduce the risk of obesity for her child in the future overlooks
the quality of life that may already affect marginalized people of color.


## METHODOLOGICAL IMPLICATIONS
## OF STUDYING "THE ENVIRONMENTS"

Examining the impact of the maternal environment on health across the
life course is methodologically challenging because the relevant environmental variables are not discrete or stable, and the underlying logics of
the RCT design cannot capture the on-the-ground realities and entangled aspects of "the environments."[27] For instance, the design of an RCT
focuses on finding linear causal associations between discrete variables
(like food or weight) and health outcomes—that is, causal correlations
between an intervention and an outcome. As I have shown, what comes to
count as the maternal environment can be oriented around the pregnant
body, behaviors, metabolism, food as calories, and the home. Different scientists will frame their environmental variables in distinct ways. Critics
may say, "Well, this is a discursive difference, not scientific." My point is
that the terms, languages, and concepts we use to define the environment
are material-semiotic.[28] The construction and definition of environmental
variables have material impacts on what is measured, the data collection
process, and intervention implementation.

   Another methodological issue is that in order to find differences
between control and experimental groups, the groups have to be composed of standard bodies or environments. Applying this notion to
pregnant bodies is complicated because assumptions about where the
environment begins and ends are precisely what is at stake when comparing control and experimental groups in an RCT. Examining behavioral
interventions and their effects on physiological or metabolic health outcomes requires not just standard bodies, but also standard households,
neighborhoods, and environmental toxin exposures. Within the scope of
epigenetics, environmental factors are entangled and enmeshed, and it is
not possible to prove one effect comes from one environmental variable,

as if it were separate from the social, experiential, and material aspects of the environment.

In other words, the RCT design is not capable of addressing the entanglement of environmental factors because it is aimed at producing causal-linear associations between discrete variables and outcomes. Focusing entirely on diet and exercise assumes that no other environmental variables are at play. Statistically, other variables are factored out to focus only on the ones that the study proposed to examine; large sample sizes allow for these types of analyses. Any variation that emerges in the dataset that cannot be explained in relation to the variable and research questions proposed is referred to as "random noise." *Random noise* is a statistical term to reference data that cannot be read by machines. It is often used to explain inconsistencies in the data that may be due to the lack of standardization among intervention delivery, since different interventionists may have different styles of presenting the intervention. This "random noise" in the data can be discounted with high statistical power. However, based on my work on the ground as an interventionist in these trials, the "random noise" can reemerge as significant when/if the intervention is applied in standard care, or in other trials.

Although it is rare for any pregnancy clinical trial to ever be reproduced, some of these interventions, like the one in the United Kingdom, were already approved and funded to be implemented in community health programs even before the trial officially concluded that the intervention was not clinically significant. The crisis of reproducibility carries implications for scientific validity. In theory, anyone should be able to reproduce these trials and obtain the same results. In reality, these results (either positive or negative) are rarely if ever confirmed. Moreover, if trials are clinically insignificant, or if studies have null findings (did not confirm their hypotheses), it becomes almost impossible to publish these results. The crisis of reproducibility and the dearth of publications on null findings carry significant implications for how health policy is made.

Even beyond the issue of the RCT method and the crisis of reproducibility, it is difficult to capture any epigenetic modifications in human models, full stop.[29] Key challenges that are cited relate to issues of sample size, uncontrollable batch effects, and unknowable factors that shape methylation at different temporal rates.[30] More research is finding that epigenetic modifications, like methylation, can change in an hour, a day, or a year.[31] Capturing this type of temporal variance is extremely challenging in one person's body, let alone in thousands of participants.

In addition, stochasticism, or random and unpredictable epigenetic modifications, further complicates the examination of epigenetics in human models. Stochasticism means that there are multiple forms that epigenetic changes can take, and that it is not possible to predetermine or predict how changes will unfold. For instance, epigenetic markings can be inherited from past generations, and if they stay intact, they may impact genetic expression or regulation. It is also possible that these markings may not stay intact, however, which may mean that epigenetic modifications are not passed on. There is also the possibility that if epigenetic markings are inherited, or if new changes are stimulated during gestation, these changes may not be activated until later in life, (this process is known as a latency). Therefore, to focus on pregnancy interventions alone ignores the indeterminate and unpredictable impacts of epigenetics in gene-regulation and expression.

## POLITICAL IMPLICATIONS
## OF EPIGENETIC FORECLOSURE

The methodological and conceptual limitations of exploring the environments in prenatal interventions illustrate a particular valence of epigenetic foreclosure. Despite the multiple forms of exposures that pregnant bodies experience, the only ones that come to matter are units of the environment that fit into lifestyle interventions of diet and exercise. Focusing on individual lifestyle interventions for chronic diseases in pregnant or nonpregnant adults is indicative of the epistemic environments that shape interpretations and applications of epigenetics. A capacious understanding of epigenetics shows us that stress, social inequalities, racism, and trauma can get into our cells and bodies. Yet selective interpretations and translations of these ideas are narrowed down to align with existing social and political climates characterized by late liberalism and capitalism.

By asking ourselves what should count as "maternal environment," we can start to reimagine a maternal environment that (re)distributes responsibility across bodies, communities, and government. At stake in the conceptualization and clinical translation of maternal environments are the treatment and care of diverse people who collectively participate in social and biological (re)production. By taking seriously the idea that how we conceive of the maternal environment shapes the interventions that are designed, future trials could employ a different approach, one that starts by asking what pregnant people and partners/family need and want. What environmental domains require intervention from their perspectives? This approach will require reconceptualizing "participants" as collaborators in clinical trials. It might also require us to rethink whether the randomized clinical trial method is the best avenue for translating environmental epigenetics.

Alternatively, it may not be necessary to invest hundreds of millions of dollars in research that focuses on changing individual bodies in order to make them resilient to toxic racist environments. More urgently, the COVID-19 pandemic makes clear that we need to shift our focus toward changing racist environments. Preventing future generations from getting sick or developing chronic illness is a moot point if people are dying in the present.[32] Through a framework of the politics of postgenomic reproduction, the more comprehensive question becomes: What kinds of environments promote healthy, safe, beings?

# 6  Prospecting Pregnancies

DATA, TIME, AND SPECULATIVE VALUE

You are scheduled for an oral glucose tolerance test. You did not eat anything after 10:00 last night, and now you're hungry. You're a little confused and nervous about this "test." You arrive early in the morning, your stomach is grumbling, and the first thing the midwife tells you is "get ready for the fasting blood draw."[1]

You hate when they have to take blood. Since the beginning of your pregnancy and enrolling in the trial, it feels like everyone's been taking your blood. When you say something to the staff about how much blood they take, the response is always the same: "Pregnant women produce 30% more blood, so there's plenty to go around."[2]

You sit down in the clinic chair, and the midwife feels for a vein with her bare hands, walking her index and middle finger across your skin. She comments, "veins are small," and continues searching. She says, "I hope this goes smoothly or else we have to take you up to the eighth floor to see the expert phlebotomist."

When the nurses and midwives take your blood, the needle usually slides in smoothly and you don't have to worry about getting poked and prodded too much. Other times they have to switch arms and tighten the rubber band around your bicep. Only once did the nurse have trouble finding the vein, and she took you to the eighth floor. You overheard her explaining to the phlebotomist, "The heavier ones have more fat around the veins, which makes it harder to find it."

You breathe heavily as the needle slides in and red fluid rushes out into the attached tube. You exhale as the needle slides out.

Once the tubes of your blood are labeled, the research assistant takes them up to the tenth floor for analysis.

The midwife says, "now it's time for breakfast" and pulls out the Lucosade.

You recognize the bottle from advertisements on the street. It's the "sport" drink that the English national team promotes. The midwife explains that it has a lot of sugar, and they use this liquid sugar for the test. She pours an amount equivalent to 75 grams of sugar into a cup.

You start to take a sip of it, but it tastes like medicine with a bitter aftertaste, masked at first by a heavy layer of syrupy sugar. You slowly take another sip.

The midwife says, "I'm going to need you to drink it a little bit quicker." Then she calls up to the eighth floor to schedule your extra ultrasound. You get a special 3D ultrasound as a reward for enrolling in the trial and completing this glucose test.

You're excited to see these images!

She repeats, "I'm going to need you to down the rest of that, because the longer you take the more you're starting to digest, which may affect the test."

Finally, you drink all of the Lucasade, and you have to wait one hour before the next blood draw. In the meantime, the midwife guides you through some survey questions.

The alarm goes off; time for the second blood draw.[3]

You finally ask, "Why did you take the blood at the beginning and at one hour? And what are these blood tests for?"[4]

The midwife explains, "Well, at the beginning is your fasting blood which shows us what your sugar levels are when you haven't eaten. Then the one-hour and two-hour bloods tell us how you're processing the sugar. We take these bloods and compare across experimental group and control group. You will not get any results from these blood tests."

You follow up by asking, "What does the test show or why do you do it?"

The midwife responds, "To see if you have diabetes."

.    .    .    .    .

This illustration, from the UK trial, is a snapshot of a glucose tolerance test at twenty-eight weeks and blood collection from a pregnant participant. Describing this process is significant for showing how pregnant bodies are prospected in a clinical trial. I frame the data collection phase of the trials as *the prospecting phase* of raw material or data for speculating on future health.[5] In the process of compiling and amassing pregnant biobits into thousands of tiny tubes that are stored in freezers, analyzed, and shared with public and private groups, one particular moment, one person, and their glucose tolerance test become obscured. In this way pregnant subjects and their biological samples, or what I term *pregnant biobits*, are made into data. Pregnant biobits include the wide variety of biological samples like blood, urine, breast milk, cord blood, and placental tissue that are collected and prepared for analysis in the pregnancy trials.[6] Documenting the mundane singular moments of data collection aims to interrupt what Donna Haraway calls corporeal fetishization, or the obfuscation of assorted relations between human and nonhuman actors that emerges in the extraction and circulation of biological material—or in the case of clinical trial implementation, pregnant biobits.[7]

I organize the ethnographic material on the processes and practices of the clinical trials' data collection phase around the themes of data and time. In so doing, I show how certain kinds of data—primarily pregnant biobits turned into biological forms of data—are made meaningful and valuable through different temporalities. The significance of my approach is twofold: it emphasizes how the future is valued over the present in everyday mundane processes, which is indexed by how the data collected in these pregnancy trials are framed as speculatively valuable beyond the efficacy of the trial in the present. This is meaningful for understanding why lifestyle intervention trials on pregnant populations are continuously funded despite the fact that existing trial results are inconclusive. The data collected at these trials can be used for other kinds of postgenomic developments and agendas, like the prediction of gestational diabetes. As I discussed in the introduction, the potential value of both biological and behavioral data is distinctly framed in a capitalist (specifically racial-surveillance biocapitalist) and postegenomic context. Biobehavioral pregnancy data are not just significant for surplus growth in the present; rather, these data are valuable for understanding future health.

Additionally, my approach reveals how certain data (like pregnant biobits that contain DNA or behavioral survey data on sleeping, eating, and physical activity) are valued differently, and still other data, such as life experiences and environmental exposures to racism and pollution, are not deemed worthy of collection in lifestyle pregnancy trials and thus remain obscured, or unseen. *The larger implication is that what data matter at what time is precisely what is at stake in the management of future health through pregnancy trials.* Put another way, the type of data (biological, behavioral, qualitative, or quantitative) that are selectively ignored or spotlighted is significant for how scientists are imagining and apprehending future health, a point I return to at the end of the chapter.

For instance, a veiled aspect of the process of transforming pregnant subjects into meaningful bits of data is race. The racialized pregnant bodies, which are initially targeted for recruitment because of their ethnic/racial identities, are decontextualized and depoliticized in the process of excavating pregnant biobits and making these bits into valuable data in trial implementation. The pregnant biobits are excavated, then processed into discrete value-neutral data that are severed from the social and racist environments within which these bodies are situated. Stratified and varied processes of valuing data relate to how race becomes invisibilized in trial implementation. In the process of trial implementation, race matters at the beginning for recruitment purposes (recall chapter 3), then disappears in the prospecting and analyzing phase of pregnant biobits. Only at the end, during the publication phase of research, are racial categories and politics looped back in to make claims about the impact of research or interventions on ethnically diverse populations. Overall, the lifestyle pregnancy trials foreground certain biological or individual behavioral data, which facilitates an erasure of racial, political, and environmental contexts.

In what follows, I explore what types of pregnancy data are collected at each trial, followed by an overview of the temporalities I examine in the pregnancy trials, like when and how the pregnant biobits are collected. Finally, I conclude with a section on predicting gestational diabetes that elucidates the speculative value of pregnant biobits. I argue that such speculations on pregnant biobits maintain racial, social, and economic implications for imagining future health.

## DATA

### Pregnant Biobits

The United States and the United Kingdom collected different types of pregnant biobits, and the process of choosing which bits to prioritize in data collection and how to collect them required coordination and standardization across trial sites. Pregnant biobits include a variety of biological matter that are excavated from pregnant bodies, then processed and prepped for storage and eventual analysis. The vignette at the beginning of the chapter illustrates the mundane encounter between a midwife researcher in the UK trial and a pregnant participant completing her twenty-eight-week assessment, which includes a glucose tolerance test and the collection of three separate blood samples. Through the process of trial implementation, pregnant subjects provide the raw materials of pregnant biobits, which are then transformed into data that can be analyzed.

In compliance with the requirements of the US consortium, the Smart-Start trial had to collect a variety of biological samples. The following were referenced at a consortium meeting: infant serum/plasma, maternal fecal sample, infant fecal sample, umbilical cord blood, umbilical cord tissue, breast milk, placental tissue, maternal hair, and maternal toe nails.[8] The SmartStart trial principal investigator (PI) was hesitant to collect all of these samples. She stated that it would be too much of a burden on the pregnant participants and discourage their participation in the trial. At first the main item collected was maternal blood. The maternal blood samples were collected only three times during the trial, which was less often than at other trial sites. The SmartStart trial administrators stated that they did not want to "burden" their participants for more blood samples, whereas other sites collected in the US consortium collected maternal bloods five to six times throughout pregnancy and postdelivery. The consent form of the SmartStart trial stated clearly that "a small amount of blood (approximately 7 teaspoons) will be taken and this blood sample will be stored for an indefinite period of time." I include this note because as the vignette reveals, there is a lot of surplus or so-called extra blood prospected from pregnant bodies during standard prenatal care and for the trial's data collection.

The SmartStart trial also agreed to collect umbilical cord blood and placental tissue if the participants consented to it. Near the end of my time with the SmartStart trial, the PI was getting phone calls from other scientists in the area who were interested in collecting maternal fecal samples. The PI eventually agreed to also start collecting maternal fecal samples, if participants consented.[9] Sharing data (particularly pregnant biobits, because they are not as accessible to scientists who do not work in reproduction or with pregnant populations) is a valuable resource and form of capital in scientific networks; thus deciding to collect maternal fecal samples brought with it opportunities for collaboration and publication.

For instance, in addition to understanding and tracing epigenetic mechanisms across mother and infant, the biobits collected from pregnant bodies were valuable for many different kinds of research agendas. As some of the collaborators of each trial explained to me, scientists really do not know that much about pregnant bodies and nutritional transfer across the placenta. The opportunity to collect blood, urine, cord blood, and placental tissue samples is a gold mine for scientists interested in organ transplantation, immunology, regenerative medicine, and stem cell research.[10]

Similarly, the United Kingdom had a consolidated list of pregnant biobits that were collected across its five sites within the larger consortium. Unlike in the United States, all the participating UK sites were on board with collecting the required pregnant biobits. The StandUp trial and other participating sites collected blood and urine four times throughout the trial. With additional consent, the trial also collected cord blood samples and placental tissue. In the UK trial the focus was placed heavily on what the staff called the "research bloods." These pregnant biobits were analyzed, compared, and processed in many different ways. I explore this further through ethnographic material from a data meeting in the UK trial.

*Behavioral versus Biological Data*

In addition to the pregnant biobits, each trial collected behavioral data. The behavioral data were valued differently than the biological and genetic data. Behavioral data included quantifiable measurements like step counts, calorie counts, and weight. They also included surveys. Each trial required extensive survey packets to be completed by each participant that included measures of anxiety, depression, smoking, pregnancy history, breastfeeding, appetite and feeding responses, and sleeping patterns, and, in the United States only, data on home environments.

The trials framed behavioral data, such as surveys, gestational weight gain, and step counts, as reliable, measurable, and quantifiable. This framing generated a quantitative analysis of the data, which stands apart from a qualitative approach. A feminist and anthropological approach to qualitative analysis prioritizes different forms of data like interviews, observations, or oral histories (which can also be framed as behavioral data). Arguably, oral histories and participant observation will produce data about a participant's environment distinct from the responses to survey questions about the living environment. The point is that how data is defined and analyzed (as behavioral or biological, quantifiable or qualitative) impacts the scientific understanding of the maternal environment. For instance, in chapter 2, individual lifestyle pregnancy studies depend on discrete measures such as body mass index (BMI) and gestational weight gain (GWG). An overdependence on such data risks foreclosing an understanding of the maternal environment, which is reductively defined as individual pregnant bodies and behaviors.

The varied amounts of time and attention invested in collecting either biological or behavioral data revealed a stratification in value and individual interest. For instance, the PI in the SmartStart trial commented that the consortium meetings on data collection (or the meetings with the collaborating sites across the United States) focused primarily on the biological sample collection.[11] She explained that the behavioral data were not as important to the other sites, because although she collected some biological data that the other trials in the consortium needed, such as placenta and cord blood, the other trial sites did not contribute behavioral data that she prioritized.

For example, the US PI was a health psychologist and an expert in designing behavioral modification interventions. Consequently, as staff delivering the intervention, we spent more time discussing the behavioral data collected from each intervention visit at the weekly mandatory staff meetings. We met and workshopped each participant and their weight, calories, and the number of meal replacement shakes consumed. In contrast, from my observations as a staff member on the SmartStart trial, there were far fewer staff meetings dedicated to the collection of the biological data.

Distinctly, the UK trial had more meetings about the biological data collection and analysis. In interviews with the staff who delivered the behavioral intervention in the United Kingdom, they explained that they did not have enough oversight or supervision in addressing emergent issues during implementation. They did not meet regularly to discuss the collection of the behavioral data or the implementation of the intervention. The PI of the UK trial, a physiologist by training, explained to me that her expertise was not in behavioral data, but rather the biological mechanisms of diabetes and obesity; thus she was more interested in the collection and analyses of the biological samples. After a data meeting at the UK trial, the UK PI stated: "I am more interested in the [DNA samples]. That is why I did this. The behavior part is not my thing, the scientific mechanistic part is vital to clinical outcomes related to metabolism, diabetes, and care." The differences in prioritizing the data in the United States and United Kingdom illustrate how the data and their collection are stratified in value depending on the scientific agenda and expertise of each PI.

The contrasting values placed on collecting behavioral and biological data at both trials reflects a context in which pregnant biobits emerge as more significant in the longer term because these data (including genetic material) are needed to illuminate mechanistic evidence for epigenetic modifications. The biobits and data prospected from pregnant bodies become what Darling and colleagues call an epistemic "hinge" for understanding epigenetic mechanisms in developmental origins of health and disease (DOHaD) research.[12] Cord blood collected at birth "may be associated with lifelong changes in the metabolic functions of the individual" that are directly connected to pregnancy and exposure in utero.[13] Láppe and Landecker note that cord blood is often the "proxy" for linking "maternal stress, infection, or obesity with changed methylation or histone acetylation in fetal and infant genomes."[14] The cord blood is a valuable pregnant biobit that materially links maternal and fetal environments.

The difference imagined across biological and behavioral data is a false dichotomy that obscures the role they play in epigenetics and racial-surveillance biocapitalism. Through epigenetics, as a theory that bridges nature and nurture "divides," the biological and behavioral material collected from pregnant bodies is entangled. Biological data linked to biomarkers like hormone levels are directly connected to experiences and behaviors. Epigenetics blurs the distinctions between biological and behavioral data. Treating each kind of data as if it is a discrete variable obfuscates the entanglement of nature/nurture or biobehavioral domains inherent to epigenetics.

I argue that in an epistemic environment of racial-surveillance biocapitalism, the behavioral and biological data together have the potential to expand forms of surveillance and to generate surplus value.[15] For instance, surveys on sleep and eating, as well as blood and urine, can be used to categorize pregnant people and parents into risky or healthy groups. The same data can be shared with private collaborators to create health products for new parents, and for predictive medicine. Overall, the ethnographic examination of pregnant biobehavioral data collection brings back down to earth the grandiose notions of big data by revealing the conditions and temporalities in which these materials are collected and made meaningful. Doing so throws into relief the politics of postgenomic reproduction. Mirroring the value placed on biological data, the rest of the chapter focuses primarily on the collection and analysis of pregnant biobits and the role that temporality plays in making these bits speculatively valuable.

TIME

*Gestational, Future, and Epigenetic*

Temporalities are characterized as benchmarks for living, birthing, and dying that are shaped by cultural values and are indexed through daily practices.[16] For example, Karl Marx explains that in order to maximize the productivity of an individual in a capitalist setting, their time and labor requires strict disciplining and a restructuring of mundane daily activities, like waking up, eating, and sleeping.[17] Such practices create a temporality of a working day that shapes an individual's living conditions. The "working day" is just one form of temporality: linear, predictable, measurable, and moving in one direction.[18]

Pregnancy trials and epigenetics, however, create distinct spectrums of temporalities. In this section I outline three important temporalities for pregnancy trials: gestational time, future time, and epigenetic time. These temporalities are indexed by certain practices and tools. For instance, each clinical trial is based on gestational time, which is a carefully crafted temporality that requires algorithms, measurements, ultrasounds, and pregnancy wheels. Another temporality is future time, which emerges as significant in the analysis phase of the trials and speculative visions for using the pregnant biobits for predictive medicine. Future time is characterized as future thinking, predictive, and anticipatory. The analysis of pregnant biobits, using biomarkers, is anticipated to tell a story about the potential risk imbued into fetal matter, infants, and future adults. I further elaborate on the role of pregnancy data in predicting future health risks by examining the case of gestational diabetes at the end of the chapter.

All three temporalities conceptually and practically coexist but do not always align. For instance, in the implementation of pregnancy trials, gestational, future, and epigenetic temporalities are not equally considered: gestational time and future time emerge as significant in the translation, dissemination, and implementation of pregnancy trials, and epigenetic time disappears from view. I emulate this dynamic by foregrounding ethnographic material of gestational time and future time and the roles each temporality plays in prospecting and making pregnant biobits speculatively valuable. The multiple temporalities that are at play reflect the disproportionate focus on the future over the present and the selective omission of nonlinear and unpredictable aspects of epigenetic time altogether.

*Epigenetic Time*

Epigenetic time is a valence of temporality that is selectively ignored in the implementation of the trial. Scientific literature characterizes epigenetic modifications to genes as unpredictable, indeterminate, and potentially transgenerational. For instance, epigenetic memory explains the endurance of modifications to DNA over time that can be unpredictably inherited across generations and can be triggered latently. Latent modifications are connected to environmental stimulations that can be inherited and epigenetically remembered from past generations and manifested in the present and/or future.[19] Due to unpredictable modifications, or stochasticism, on parts of the DNA, it is possible that these changes can actually be inherited across generations and be triggered later in life. Drawing on these descriptions of epigenetics, I frame epigenetic time as unpredictable, indeterminate, and transgenerational.

Landecker and Panofsky highlight that "stochasticism [. . .] is an essential *but often overlooked* caveat in discussing cause and effect in epigenetics."[20] From my ethnographic research, the translation of epigenetic theories in evidence-based medicine selectively ignores notions of epigenetic temporality like stochasticism and latency. This is a kind of epigenetic foreclosure: drawing on epigenetics and DOHaD theories to justify interventions now, in the hopes of securing future health, obscures the reality of nonlinear and unpredictable aspects of epigenetic mechanisms. The transgenerational component of epigenetic time is popularly imagined as the transgenerational inheritance of trauma, or the impacts on health that are accumulate over time and directly connected to violent exposures like slavery.[21] Interpretations of transgenerational inheritance are racialized and can be used to justify the disinvestment in public welfare due to assumptions around individual predisposition, an aspect that I return to in the conclusion.

In clinical trials, epigenetic time is selectively ignored, in part because it is challenging to capture in evidence-based medicine and in human models. Most of what is understood about the temporality of epigenetics is based on observational or animal studies.[22] As I discussed in the previous chapter, the clinical trial method in human models cannot assess, measure, or predict the intersecting forms of environmental exposure that occur(ed) across the past and present. Epigenetic time makes it challenging for the randomized clinical trial (RCT) method to capture epigenetic modifications.

In addition, as I explore further, the pregnant biobits and biomarker analyses are searching for the presence of an activated biological sign or signal that is connected to an exposure in the uterine, fetal, or maternal environment, not the absence of a sign. Due to stochasticism, it is possible to *not* inherent or manifest epigenetic modifications from the past.

Yet current trial and study designs are unable to capture the potential to *not* inherent epigenetic modifications. Instead, the spotlight is placed on how pregnancy as a critical period can stimulate epigenetic modifications that are risky for future generations. Such a hyperfocus on pregnancy and pregnant bodies' behaviors makes pregnant biobits valuable for speculating on future health, but not necessarily on the present.

If epigenetic time is comprehensively considered in the design of pregnancy trials, the question remains: *Is pregnancy the most effective time to intervene in future health?* For instance, a pregnant person in the present, whether they are obese or not, could have already inherited epigenetic modifications from their grandparents or great-grandparents, which could have already affected fetal development in the present and future. Through epigenetics, the risk of modifications can simultaneously be inherited from the past, experienced in the present, and potentially passed on to future generations, or it may remain latent. Thus, it becomes challenging to assess whether exposures in the past or present are linked to future health outcomes. I emphasize that the selective ignorance of the unpredictable and indeterminate aspects of epigenetics enables a prioritization of the future.

Next, I ethnographically explore how gestational and future time make pregnant biobits meaningful through clinical trial implementation. Doing so reveals the misalignment that exists between the grandiose ways in which epigenetics is imagined to fundamentally change conceptualizations of health, disease, and inheritance, and how scientists are actually applying epigenetic ideas on the ground.

## MAKING GESTATIONAL TIME

At my desk in the US SmartStart trial, I had to learn how to use a "pregnancy wheel," a tool used to determine gestational age based on the last menstrual period (LMP) (see figure 4).

The wheel is made of two circles; the internal circle rotates while the outer circle with the months stays stationary. To calculate gestational age and estimated delivery date, one is supposed to align the arrow labeled "first day of last period" with the date provided by the pregnant

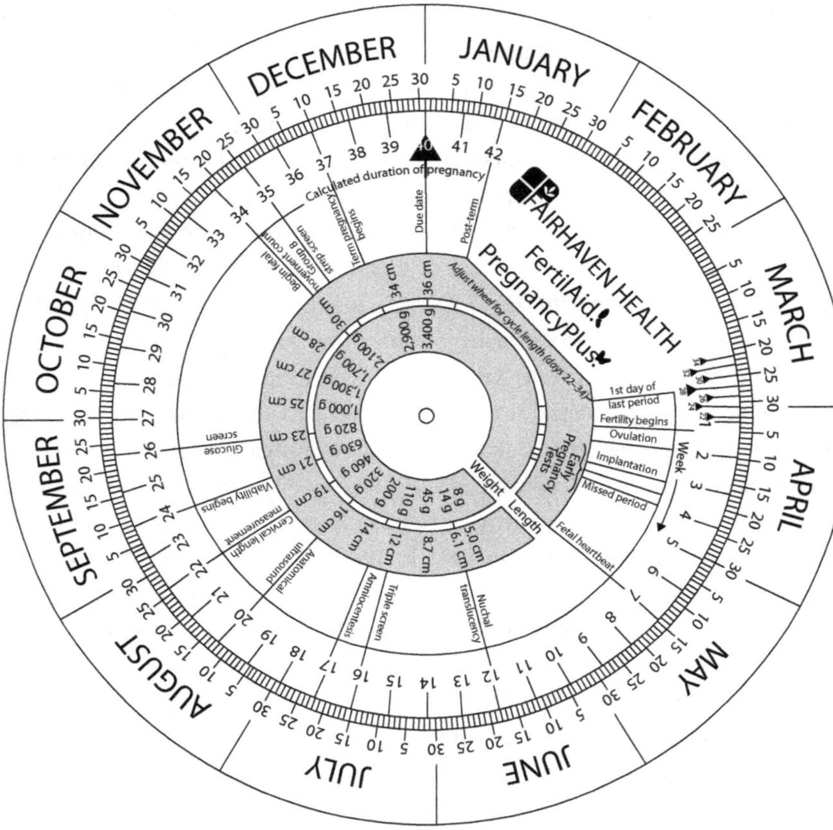

*Figure 4.* Pregnancy wheel for measuring gestational age. Fairhaven Health, FertilAid/
PregnancyPlus.

participant. The other arrow on the other end of the inner circle points
to the "estimated due date" positioned on the outer circle. This time-
structuring tool produced what was understood as gestational time and
also helped determine the inclusion of participants in the trial. During
my training, I built an intimate relationship with three terms: LMP, estimated date of delivery, and gestational age (GA). These indicators of time
created a whole new world for me (as well as for the trial) that depended
on different methods of counting dates and different reference points for
when to start and stop counting time.

The first time I used this wheel was during a phone screening for eligibility. I would call up the potential participants and say: "Hello, this is Natali calling from the SmartStart trial, we received your information from clinic X, do you want to learn more about this trial?" About five minutes into the call, after we had assessed race, ethnicity, age, and BMI, I would then ask: "When was your last menstrual period? How sure are you that this is the date of your LMP? Are you extremely sure? Somewhat sure? Or not at all sure?" In my phone screening script these instructions were in boldface: "Use gestational wheel to calculate weeks pregnant based on LMP."

Gestational time (or gestational age as it was labeled in the US protocol) played a significant role in dictating and structuring the implementation of each trial. It determined when to recruit, whom to recruit, when to intervene, and when to collect data. Gestational time had to be calculated carefully and vigilantly. This was not done in the same way in the United States as in the United Kingdom. Here I mainly focus on how the US trial crafted gestational time, because in the United Kingdom the administrators of the trial y were able to recruit straight from the first prenatal appointment. They also had access to everyone's records through the National Health Service (NHS). Therefore, the individual staff members did not have to confirm gestational age through other means.

In the United States, the GA was based on the pregnant participant's date of LMP and an official ultrasound. The protocol in the United States provided specific steps for how to calculate the GA using the LMP or ultrasound data from first prenatal appointment. For instance, the protocol stated that the "gestational age at randomization [for inclusion in the study could be] no earlier than 9 weeks 0 days and no later than 15 weeks 6 days based on an algorithm that compares the LMP date and data from earliest ultrasound."[23] The SmartStart trial was required by the National Institutes of Health (NIH) consortium to also use an extra measure for calculating GA. This was an algorithm that we had access to via the trial website, which used the LMP to calculate a GA. The algorithm was used to ensure the standard practice of finding GA across all participating sites.

In addition, the US consortium had a calculation to determine BMI based on time of gestation. For instance, the trial protocol explained that one or two pounds could be subtracted from the initial weight measurement of the participants if it was completed at fifteen weeks versus fourteen weeks' gestation.[24] In this way, gestational time also indicated how many pounds were attributed to maternal weight, which was separate from the weight due to fetal development and placental growth in the first trimester. The BMI depended on the weight at the "start" of pregnancy. I place "start" in scare quotes because the moment of starting pregnancy is precisely what is at stake and produced through the tools, practices, and methods of assessing GA.

It was not an easy process to learn how to count gestational time or age. Maintaining gestational time required vigilant standards, measurements, algorithms, tools, accountability, and audit practices. For instance, I was trained to carefully check at the first screening what the LMP date was and whether the pregnant participant was certain about that date, then I had to match the self-reported LMP with the official ultrasound results. Making gestational time required multiple checks and balances. I had to confirm the "start" of pregnancy in three different ways: by using the "pregnancy wheel"; by using the algorithm provided by the consortium; and by documenting the prenatal ultrasound, which we had to gather from the participants' doctors' offices.

In the ultrasound, the GA is assessed by measuring the diameter of the "gestational sac" and the crown-rump length (CRL). For example, if the CRL is 10 mm, then the GA is estimated to be seven weeks. This measurement is time sensitive, because the more time passes, the more variable the fetal growth becomes. The ultrasound can best assess the size of the fetus during early pregnancy, when the size of the fetus is more consistent across different bodies.[25] The smallest "sac size that can be clearly distinguished by an [ultrasound] is 2–3 mm, which corresponds to a gestational age of about 32 to 33 days."[26]

There is no clear consensus on whether GA is better assessed using the first day of the last menstrual period or by measuring the CRL from the ultrasound, but both are used to ensure accuracy. When I asked one of the staff members why it was so complicated to find a GA she responded, "You have to find a standard way, everyone is different. [In] my last pregnancy my LMP said I was nine weeks pregnant but when I did the first ultrasound it said five weeks; they have to find a standard way."[27]

The methods and practices designed to maintain and confirm consistent gestational time required institutional surveillance and documentation. Only once did I observe a documented mistake in which the randomization occurred without confirmation by the ultrasound. This mistake required the submission of a written report to record and describe the "protocol violation." To prevent human error in calculating gestational time, staff were repeatedly reminded of the protocol at the weekly staff meetings and the monthly consortium meetings. After one such meeting the project coordinator expressed frustration with how much time was spent talking about the protocol for estimating GA.[28]

Gestational time is not only fundamental to structuring and organizing the trial implementation, standards, and procedures; it is also vital in determining *when* pregnant biobits are collected. For instance, as the introductory vignette illustrates, the participants had their glucose tolerance test done at twenty-eight weeks, which required fasting bloods. These bloods were analyzed for triglycerides and different kinds of cholesterols that assessed the maternal lipids and glucose metabolism.[29] Certain tests reflect particular kinds of connections and mechanisms if they are captured *at certain times*. The maternal blood samples at twenty-eight weeks are a snapshot of the maternal metabolic environment, which are then connected to the results of cord blood and placental tissue samples collected from the infants at delivery. In this way, the gestational time of data collection and their analyses are meaningful and valuable in *telling a story* about the risks of metabolic syndrome across the maternal and infant bodies. The collection of samples in the present is fundamental to producing knowledge about future risk.

However, what is actually connected are biochemical materials in maternal blood and materials in the cord blood of the infant. These materials are extrapolated to tell a larger story about maternal behavior, fetal and infant development, and the environments that shape the biobits that are analyzed. There are other ways to tell this story of maternal, fetal, and infant environments, and those data were primarily gathered through surveys on food and sleep patterns. There are still other ways to tell this story, which might include recording oral histories of the pregnant participants, but these data were not collected.

In a broader context outside the trial implementation, the difference between week 9 and week 12 of a pregnancy, if improperly assessed, is a significant temporal difference with political and bodily consequences. For instance, different states across the United States allow or do not allow certain kinds of abortion at certain gestational times. Gestational time is also fundamental to prenatal and genetic testing and to fetal surgery and neonatal intensive care research.[30] The time that it takes a fetus to develop its vital organs, or what medical experts deem as vital, which presently includes the membranes and structure of the lungs, is also time dependent. Currently, emergency-induced births can occur around twenty-four weeks' gestation.[31] It is at this time that medical experts have the knowledge and technology to assist in the preterm birth of a fetus. In these particular cases, definitions of gestational time are shaped by medical knowledge, technological capacity, and social values to apprehend fetal "birth" and "death."

## FUTURE TIME: PREGNANT BIOBITS, DATA, AND BIOMARKERS

> In July, we will start the first studies in the first children [from the StandUp trial] and we will be looking at them as they grow up. [. . .] We have been funded very well by MRC and European Union consortium. [. . .] And actually I'm going back to my love of biomarkers now and we have a lot of money to look at the biomarkers in the mother which might be influencing the developing fetus and therefore we'll get a handle on those associations, and we are using lots of fancy techniques [and] technology to look at that.

This excerpt comes from a public speech by the PI of the StandUp trial at a health research conference in the United Kingdom. The conference was well funded and included scholars from elite institutions, medical practitioners, and government officials like the director of the NHS. As the PI suggests, the scientific practice of collecting and analyzing biomarkers is vital for understanding the molecular modifications in fetal development. As she also states, she is very interested in following up on the children of the pregnant participants. Through epigenetics the children could potentially inherit the biochemical changes that may occur as a result of being exposed to a nutritional intervention in utero. Biomarkers are used to trace an exposure in the uterine environment to an effect or genetic expression in the offspring. Biomarkers are one way to capture and assess these changes in relation to outcomes, and they require pregnant biobits that are primed and stored for future analysis.

In clinical research, biomarkers are defined in multiple ways; they can be "any substance [including blood pressure, pulse, blood fluid and other tissues], structure, or process that can be measured in the body or its products and influence or predict the incidence of outcome or disease."[32] Biomarkers can illustrate biological processes that are assumed to be objective and measurable.[33] Essentially, biomarkers are signs collected from biological matter and connected to outcomes and clinical endpoints.[34]

Biomarkers are indexical of speculative science and medicine. I frame biomarkers as an instantiation of future time. They represent the belief in prediction and prevention by targeting the individual and focusing on the molecular scale to trace mechanisms.[35] Scholars caution against the belief that biomarkers will be the solution to predicting or diagnosing disease because there is still a degree of uncertainty in tracing gene expression across the life span.[36] For instance, in Alzheimer's research the focus on biomarkers is based on the idea that Alzheimer's can be detected, at the molecular level, in individuals twenty years prior to their experiencing any symptoms. To be clear, biomarkers can detect the presence of a biological sign that is associated with a health outcome (and even these associations are unstable), but they do not offer a cure or means of prevention. Focusing on identifying molecular biomarkers in individuals that can *detect* disease potential, or a potential risk of disease, draws resources away from treating and investing in environments of care for people who are experiencing Alzheimer's in the present.[37]

To study the biomarkers and DNA of the pregnant participants and children, the StandUp trial has a team dedicated to biological data analysis.[38] One of the key people on this team is Connie. Connie, a four-year veteran on the UK trial, is in charge of monitoring and organizing all the pregnant biobits and prepping and storing them for analysis. Connie played a key role in the process of transforming raw material of pregnant biobits into analyzable data.[39] On one of the days I shadowed Connie, I asked her about the kinds of analyses the trial could do with the pregnant biological data.

Connie responded, "They could do infinite analyses on these samples." By "infinite" Connie implied that if money was not an obstacle, the PI and her colleagues could do a limitless number of analyses.

For instance, each participant has four sets of research bloods drawn, and each set requires four different kinds of preparation with different serums and DNA extractions. Each set is extracted at three different times during gestation and once postpartum, reflecting different changes in the body. The blood processing results in fifty vials that need to be frozen and stored until they are sent to another site for analysis. A key problem for Connie was making sure that the five different participating sites are all collecting, prepping, recording, and freezing the DNA samples in the same way. To ensure standardized preparation and storage, the PI decided to have all biosamples from all the sites sent to the main site, where Connie was in charge of organizing them. With fifteen hundred participants and fifty vials each for four sets of research bloods, there are over 300,000 DNA and urine vials that Connie needs to record and properly store, on top of the daily bloods that come in for processing.[40]

· · · · ·

In early May 2014 I attended a "data conference call" at the StandUp trial with the PI, Connie, and a few other collaborators to discuss plans for the data analysis. The trial was awarded around £1 million for analyzing the biological data, which did not include the behavioral data from the intervention delivery. At the beginning of the meeting the PI asked to have all baseline DNA and urine analysis by June 2014, since the trial had finally reached the recruitment goal of fifteen hundred just that week. After the group discussed some of the key logistical challenges to meeting this goal (such as freezer space and staff to process the samples), they began making a list of key biomarkers that they wanted to examine.

At the meeting, the PI briefly described a list of biochemical substances to examine in the pregnant participants' blood and urine samples: feraline and cytokines. Feraline is linked to liver disease in nonpregnant adults, but nothing is known about the effects of feraline in pregnant adults. The group decided to examine feraline in the pregnant women's DNA because of its publication value, or impact on the scientific community, in addressing pregnant adults. Cytokines are linked to neonatal adiposity, or how much fat an infant is born with, and they were approved for analysis because of their potential to indicate fetal programming during gestation.

The PI commented that she was interested in examining cytokines in the blood of the pregnant women because the animal models suggest that cytokines may cause extra fat deposits in the new infants. Since another large clinical trial was also testing for cytokines, she decided to include it but specified that the test should be done on the bloods collected at twenty-eight weeks' gestation (during the glucose tolerance test; recall the vignette) to see the influence on fetal development. Cytokines tell a particular kind of story *if* they are collected at twenty-eight weeks.

The purpose of testing the maternal bloods for cytokines is that if they are present at a particular time of gestation, they can influence fetal development in the uterus, resulting in an infant with high fat deposits, or adiposity, which is linked to obesity in adulthood. Cytokines are associated with highly processed foods and foods high in sugar, salt, and saturated fats. Only recently have studies shown a connection between cytokines, nutrition, and epigenetic modifications.[41] Their nascent role as indicators of nutritional and metabolic health across the maternal and fetal environments remains inconclusive, which is why studies need pregnant biobits to confirm cytokines as a reliable biomarker. Cytokines may detect adiposity in infants, which is then extrapolated to evaluate the risk of metabolic syndromes like obesity and diabetes. However, cytokines do not illustrate the environmental contexts that make their presence possible in pregnant bodies and infants.

As these biobits are transformed from blood, to vials, to analyzable data, the results will show differences in the data—for example, some samples may have higher forms of cholesterol, cortisol, or cytokines—*the presence of these biochemical substances in the pregnant biobits is disconnected from the environments that produced them.* In this form, we cannot explicitly see the connection between samples that have higher cortisol levels because of long-term exposure to stress and racist environments. At the analysis phase, these biobits are assessed for the presence/absence or levels of biochemicals, and not whether the samples came from particular toxic environments.

Black feminism emphasizes that environmental contexts are not only structured by gender, race, and power, but also shape material conditions of *exposure* for individuals.[42] The emphasis and analysis of particular types of data, is political: these pregnant biobits become the main focus for the production of future predictive medicines, and the environments and living conditions of the pregnant participants in the present disappear from view. Even though it is well documented that stress and racism can make us sick, the ways in which epigenetics is applied and translated in lifestyle pregnancy trials continue to ignore processes of racism and racialization. My point is that the type of data and the environmental context that is selectively ignored or spotlighted are precisely what is at stake when scientists are imagining and apprehending future health.

Once the data meeting was over, and the group had finalized the list of biomarkers they were interested in examining across the pregnant women and their children, the PI exclaimed, "This is such a fascinating time to be doing this. I am so excited!" She went on to say that the more tests they did, the more reliable their results would be. The PI stated that with this (biological) data they would publish a "paper on [diabetes] in pregnancy, [that] will be ahead of the game, addressing the issue of the chicken or egg. Is insulin resistance a predictor for diabetes?" The statement was related to a larger research agenda for the UK PI, in which she was trying to find a way to predict gestational diabetes before women develop it. If there is a common biomarker (or biological sign) from the biological samples and tests of participants who developed GDM, then perhaps pregnant people can be tested for this sign early on in their pregnancies and be channeled into a different kind of treatment and prenatal care to prevent the development of GDM. This is a specific manifestation of how biomarkers are imagined to be the predictors of future disease and how this information might change prenatal care. The pregnant bodies that provide the raw materials used to make these future imaginaries possible are also the sites on which these future interventions are tested.

## SPECULATIVE VALUE: INVESTING IN FUTURE HEALTH

In a public lecture on the research significance of the StandUp trial, the PI explained that "the maternal metabolic state plays a very important role in the *future risk of disease* in the developing child" (my emphasis).[43] This is the type of anticipatory discourse that justifies nutritional interventions during pregnancy in the present. The underlying message is that intervening in and studying the maternal environment in the present can prevent future disease, *and* that doing so is financially beneficial to the state. In one of the preliminary publications of the StandUp trial in 2011, the PI further explains that the research and its intervention have the potential to save the NHS millions of pounds in health-care costs related to childhood and adult obesity and diabetes care. The financial logic is that treating diabetes now is expensive, but implementing lifestyle interventions in the hopes of preventing obesity and diabetes in the future is cheaper. Exploring the case of GDM further illuminates how contemporary pregnancy trials are linked to speculative value.

GDM is the development of diabetes during pregnancy that can "go away" after birth. It is associated with health risks for infants, like macrosomia, or "fat babies," and for pregnant people it is associated with the development of type 2 diabetes later in life.[44] The UK trial was well situated to explore GDM. Recall the glucose tolerance test, illustrated at the beginning of the chapter. Unlike some of the other tests and samples, which were optional for participants to consent to, the glucose tolerance test was unique in that it was a key requirement for trial participation. The data generated from the glucose tolerance test can be used for the development of a biomarker to predict GDM.[45]

In addition, the UK trial specifically included a clause in the consent form to allow collaboration and data sharing with private companies, which means that the data prospected from publicly funded trials can be used by private companies to make predictive medical products like biomarkers. The global health market related to obesity and diabetes is valued at billions of dollars due to the high rates of prevalence across the United States and United Kingdom.[46] A biomarker that could index who will get GDM *before* it manifests carries potential value for reducing health-care costs in the future, as well as for the creation of a profitable health product that can be sold to a broader market.

In addition, even without a predictive biomarker, treating "at risk" groups before they develop or are diagnosed with GDM is also framed as speculatively valuable. For instance, a recent assessment of GDM in the United Kingdom states: "The shift from identifying women at future risk of type 2 diabetes, to trying to predict risk of perinatal and longer-term ill-health outcomes in the infants of women who have had GDM, has prompted changes to diagnostic criteria. Criteria with lower thresholds will identify more women at risk, thus increasing prevalence and if treatment strategies remain unchanged, costs will increase. However, providing treatment to more women may reduce the risk of perinatal and longer-term ill health, potentially saving money for the UK NHS (and the individual)."[47]

The assessment claims that treating women for GDM in the present, regardless of diagnosis, is *cheaper* in the long term, based on anticipated or projected health-care costs for rates of diabetes in the future. This approach to preventing GDM was already integrated into pregnancy trials when I was completing fieldwork. During the StandUp trial implementation, another NHS-affiliated pregnancy trial was testing a pregnancy intervention that included giving ethnically diverse obese pregnant participants metformin, a drug aimed at treating diabetes, at the start of their pregnancy, before they were even tested for or diagnosed with GDM. The aim was to treat every obese pregnant person as if they already had GDM.

While the prediction and prevention of GDM and diabetes is a worthy investment, the issue here is: What kinds of interventions are deemed valuable to invest in, and who decides? Speculative and future-oriented approaches that shape contemporary pregnancy trials promote individually-based lifestyle and pharmaceutical interventions. However, such interventions do not address the conditions and environments that perpetuate health risks in the first place. These interventions do not address the systemic and structural disparities that strongly shape maternal and infant health outcomes. Unequal social and economic conditions are linked to systemic racism and index racist environments. Socially unjust and racist environments place poor Black, Brown, and Indigenous people continuously at risk. In other words, if a Black woman in the United States or United Kingdom takes a pill during pregnancy to reduce her chances of developing GDM, she may not develop GDM during pregnancy, but she is still at risk of developing metabolic syndromes after pregnancy, and her Black child still has a higher chance of premature birth and death compared to white infants, regardless of socioeconomic status. Targeting ethnically diverse people for individual lifestyle or pharmaceutical interventions does not disrupt root causes of health disparities.

Importantly, we *already have* a key indicator for the development of GDM: houseless pregnancies. Although the risk of GDM is often associated with high BMI, more research shows that BMI is not the strongest indicator of GDM.[48] Factors like unstable housing are strongly associated with the development of GDM, but clinical trials continue to focus on BMI and associated biomarkers like cytokines that link maternal weight and behaviors with infant adiposity.[49] The research is limited because there are few if any relevant grants. The funding trends for understanding obesity during pregnancy are geared toward individual lifestyle interventions or silver bullet cures like a pill.[50]

The point here is that the integration of new science in contemporary pregnancy trials is strongly influenced by capitalist speculative logics, rather than by the health needs of the dominant majority and the most vulnerable in the present. Addressing unstable housing does not require the "life-saving" innovations of science and technology industries. As such, it is not an attractive research area to fund in a postgenomic era that remains keenly focused on predictive, speculative, and profitable cures.[51] A key consequence of the overdetermined individualistic and future focus of contemporary prenatal interventions is that it draws scarce resources away from addressing present-day, structural issues of health, which negatively impacts the very same communities that are targeted for intervention.

Those who are in the position to speculate on future remedies are not the ones who suffer the consequences of failed predictions.[52] That is, the people in charge of investing millions of dollars of public money in clinical trials in the hope of finding a cure for obesity or diabetes will not suffer the consequences of an emaciated public health-care system. Throughout my field work in the United Kingdom, I witnessed news coverage on budget cuts to the NHS that would dismantle entire sections of healthcare and cut down on staff. The dismantling of and disinvestment in the public health-care infrastructure is simultaneously occurring alongside the implementation of speculative interventions on future health. By illuminating both the narrowed visions that exist *and* the solutions we already know, the case of GDM spotlights the ambiguity of twenty-first-century politics of postgenomic reproduction.

.     .     .     .     .

Exploring processes of data collection in ongoing pregnancy trials elucidates how certain data are made speculatively valuable in and through
time. In addition, tracing mundane processes of clinical trial data collection reveals how certain data (like biological data or quantifiable individually based data) are deemed worthy of collection and analysis, while other
types of data, like oral histories that contextualize people's social and cultural environments, are not as meaningful in lifestyle intervention trials.
How and what data matter in pregnancy trials that draw on epigenetics
to anticipate or predict future health has racial, political, and economic
consequences. Drawing on Black Marxist and feminist frameworks, an
ethnographic examination of pregnancy trials reveals the entangled forms
of racial-surveillance biocapitalism that inform how pregnant biobits are
prospected and how they are made valuable for future health markets.

This chapter has emphasized a few main points. First, in the current
epistemic environment, the biobehavioral data collected in pregnancy
trials are made speculatively valuable through clinical trial implementation. The emphasis on future time indexes the ways in which predicting future health is treasured over the investment in present health. In
addition, prioritizing pregnant biobits in order to develop biomarkers to
predict who may or may not develop gestational diabetes draws attention and resources away from addressing racist environments that already
shape current health disparities. Put another way, the pregnant biobits
that are prospected and made measurable and comparable are taken out
of embodied contexts and thus enable the obfuscation of the racial politics that shape the environments within which these bodies and pregnant
biobits are situated. Finally, this chapter has demonstrated that the type of
data collected, the selective environments, and the temporalities brought
into focus in pregnancy trials are directly connected to how epigenetic
science is used to apprehend and intervene in future health issues. Such
dynamics help characterize the epistemic environments that limit scientific creativity in the present.

# Conclusion

THE AFTERBIRTH OF FORECLOSURE

Together, the fields of epigenetics and developmental origins of health and disease (DOHaD) uniquely characterize the politics of postgenomic reproduction in the twenty-first century. In theory, epigenetic science offers an exciting and capacious biosocial framework of inheritance and biological malleability. However, after more than half a century of pregnancy studies on diet and exercise, and the past two decades of integrating epigenetics and DOHaD research into maternal health, contemporary pregnancy trials continue to design familiar lifestyle interventions. While epigenetic theories maintain there are capacious forms of "the environment" that can impact genetic expression, the maternal environment in contemporary pregnancy trials remains foreclosed to only individual pregnant bodies and behaviors; not even "fathers" or partners are substantially integrated into the interventions. The epistemic environments explored in the previous chapters show how interpretations of new science are limited and driven by underlying logics. For instance, within twenty-first-century capitalist and liberal contexts, even inconclusive pregnancy trials still produce valuable amounts of biobehavioral data that can be used for the development of medical products such as glucose-moderating shakes or biomarkers aimed at predicting gestational diabetes. Such products and aims are speculative and not immediately relevant to the present lives and experiences of pregnant people who participate in these trials.

One key issue with the integration of epigenetic theories in contemporary pregnancy trials is that the randomized clinical trial (RCT) trial design cannot capture the unpredictable and nonlinear aspects of epigenetic modifications.[1] Epigenetic science maintains that environmental exposures during gestation *may or may not* impact infant and child health outcomes, these modifications are not predictable, and they are not controllable by individual choices. While new scientific frameworks in postgenomics provide expansive theories and technologies for understanding exposure and malleability of biology beyond pregnancy, pregnancy trials and interventions do not reflect the creative possibilities in reframing current health problems and solutions. The focus on individual-level interventions, an overemphasis on causal-linear methodologies aimed at prediction, and the obfuscation of systemic racism in various scales of scientific knowledge production constrain the applications and interpretations of epigenetics and DOHaD research in contemporary pregnancy studies.

This process of flattening, or reducing epigenetic potentials to fit within the existing socioeconomic and scientific methodologies, is what I call *epigenetic foreclosure*. Epigenetic foreclosure illustrates how epistemic environments can hinder scientific creativity and also reveals areas for growth. I emphasize that epigenetic foreclosure is not inevitable. Tracing foreclosure spotlights both the closures and openings for different interpretations and alternative imaginations of future health.

## AFTERBIRTH OF FORECLOSURE: INCONCLUSIVE FINDINGS

The inconclusive results of individually-based lifestyle interventions in contemporary pregnancy trials are one valence of epigenetic foreclosure. Both the US and UK pregnancy trials found that their intervention was clinically insignificant. That is, the lifestyle intervention did not significantly improve pregnancies or offspring health. Failure is not a welcome finding in the scientific community. Top journals do not publish negative findings, which places people's careers and rapport with funders in a precarious position.[2] The stigma around inconclusive results forces scientists to find something in the data that is meaningful, running different statistical tests in a variety of ways to shape analyses into significant results.

Even trials that find clinically significant differences in their results are often not able to reproduce their findings because funding is limited or nonexistent for replication studies. This is currently called the "reproducibility crisis" in science.[3] The journal *Nature* has published a variety of editorials on the issue of reproducibility, and in 2016 the editors explained that scientists experience high pressure to publish and will selectively report their findings to diminish or obscure negative results.[4] The "reproducibility crisis" reveals issues of politics and incentives that challenge the corroboration of science. Trials that test lifestyle interventions on pregnant people are rarely if ever reproduced.[5] There are comparable trials that scientists use as guides, but they are not designed to replicate results. The failure to reproduce science is at the center of an epistemic predicament and illustrates another iteration of foreclosure.

This "crisis" of knowledge production also provides fertile ground for the examination of postgenomic reproductive politics. As I explained in the introduction, postgenomic reproductive politics is a framework and feminist tool that can be used to examine the indefinite unfolding of epigenetics as a reproductive science. In a broader sense, the framework can be applied to examine the limits and possibility of (re)producing scientific knowledge production. Through this framework I examine how the United States and United Kingdom interpret their findings and how these results impact the development of future projects.

## US and UK Findings

In the summer of 2019 I returned to the SmartStart trial site on the West Coast of the United States to visit the principal investigator (PI) and staff. While there, the PI and I reviewed the recent publications on the trial's individual results as well as the consortium results. I asked her: "What do we now know about lifestyle interventions during pregnancy?" She replied: "I think we would have to say that there is no evidence that lifestyle interventions during pregnancy can reduce offspring risk of obesity during the first year. Lifestyle interventions may reduce risk of C-sections, can reduce about two kilos [of gestational weight gain], may have benefits on other cardiovascular disease factors although the findings are mixed." Ultimately, the results from the individual trial as well as the consortium found no conclusive evidence that nutritional or behavioral interventions during pregnancy impact children's risk of obesity; the PI included the caveat "during the first year." This caveat suggests that long-term studies on the children are necessary to confirm the absence of potential impacts.[6] Cleaving spaces to continue doing research is also part of interpreting results; inconclusive results imply that some of the same questions still need answers. The afterbirth of foreclosure requires mediation or reinterpretation of results.

The epigenetic literature that the PI used in support of the US trial claimed that reducing gestational weight gain (GWG) could impact the offspring's health. However, the findings show that prenatal nutrition or behavioral interventions only reduce GWG by 1.6 kilos. This modest reduction in GWG does not impact the offspring's health. At the one-year postpartum follow-up with the mothers, the study found that this effect (1.6 kilo reduction) was sustained, but mainly among participants with high income and education.

Considering these results, I asked the PI if funders still support trials that focus on GWG in the United States. She said "funders now would say that GWG is not innovative to test in prenatal trials, it was innovative before because the epigenetic research supported it but it's not now." Since the Institute of Medicine (IOM) reports usually come out every ten years, I followed up with a hypothetical question on whether the next report on maternal nutrition and weight would still focus on GWG as a major indicator, variable, and outcome measurement. The PI gave a response that left the door open for more studies to include GWG: "We still don't know what health impacts GWG has on class II or III morbidly obese pregnant women." The implication of this is that "morbidly obese" pregnant bodies are the new target for understanding the impact of GWG.[7] In fact, since this interview, the National Academy of Medicine published its 2020 review of the current literature, which supports the use of GWG as a standard for maternal weight and nutrition.[8] The focus on testing different bodies, but not different standards, reflects the effort to maintain

GWG as medically relevant in the United States. This is also the afterbirth of foreclosure.

The SmartStart trial included a small ancillary study that targeted the partners of the pregnant participants. The partner's weight was measured at the beginning and end to see if there was any "ripple effect" from the pregnant person's participation in the intervention group. This study did not find any effect on weight among the partners of the participants. It was small, since most of the partners declined to participate in it. Most, if not all, of the trial's resources, staff, and time were invested in the pregnant participants and not necessarily their partners. From the lens of postgenomic reproductive politics, the effort to include partners reflects an intention to recognize the role of paternal effects in fetal and infant development, but the medical infrastructure and social value do not yet exist to significantly integrate partners into pregnancy trials.

The UK results were similar to those from the United States and other comparable trials in Australia and Ireland.[9] The StandUp UK trial did not find any clinically significant differences in pregnancy or infant health outcomes between the intervention and control groups. Technically, the intervention did not achieve its goal of reducing gestational diabetes mellitus (GDM) or pregnancy complications. Although the trial administrators did not find any clinically significant results for the infants, they are currently following up with the children at three years of age to see if there are any new developments in their health outcomes. However, one finding emerged from the UK trial: the pregnant participants in the intervention group gained on average one kilo less than the control group. As in the US trial, this difference in weight gain had no impact on pregnancy complications or offspring health.

I want to emphasize the one kilo weight reduction, because the participants and staff members across both intervention groups worked really hard for this weight reduction. The difference of one kilo required thousands of pregnant bodies, staff, labor, time, resources, money, and health infrastructure. However, as I showed in chapter 6, the collection of pregnant biobehavioral data that also depend on this infrastructure is deemed valuable for the future regardless of the inconclusive results.

A key finding that emerged across both the US and UK trials was that the intervention had no significant impact across the racial/ethnic groups in each respective trial. When I interviewed the PI in the US trial, I asked her about this issue, and she reiterated the question in her own words: "Why are the Hispanic women more likely to have long-term health problems and obesity compared to non-Hispanic women, and yet the interventions are equally [in]effective?" She responded to her own phrasing of the question: "My hunch is [that] after delivery we withdraw the intervention, the women with less income, they go back to an environment that is more obesity-promoting than the women with more education and more income. The women who are more educated can apply the tools we gave them, but we didn't see any weight change postpartum after our study, but maybe that would unravel more long-term. Acculturation didn't moderate our intervention; [that is,] the women who are more acculturated are at greater risk of unhealthy eating. I think it's economic." The PI's response to the fact that health outcomes among "Hispanics" are worse than among non-Hispanic white women focused on economics, class, and education. This interpretation echoes epidemiological and public health discourses that often conflate race and class.[10] The lifestyle intervention was designed to target individual's behaviors, which are challenging to sustain in inequitable contexts.

The UK results also found that the intervention did not have a significant impact across "ethnic origin," which was defined as "White, Black, Asian, and Other." In a publication of the StandUp trial results, the authors stated that Black women have higher rates of obesity in their urban context, and that this is related to the prevalence of low socioeconomic status among these groups. The PI and her coauthors emphasized socioeconomic status to explain the disparity of health outcomes. Similar to the US PI, this statement also conflates class with race. It is no coincidence that Black populations in the United Kingdom have lower socioeconomic status than white British citizens and also have worse health outcomes.[11] This correlation is an enduring symptom of racism. Why is racism so hard to find in research results, yet inequities in socioeconomic status are consistently used to explain persistent health disparities across race?

Individual lifestyle interventions assume that everyone has the same "healthy" options from which to choose. Yet pregnant people classified as "high risk" for diabetes and obesity are embedded in racist and poorly resourced environments that make it nearly impossible to sustain lifestyle interventions during or after the trials. While it is challenging to examine the intersectionality of race and class in a clinical trial design, a critical race and feminist lens is vital for examining how processes of racism and inequality matter in ways that are not reducible to class or education alone.

For instance, bodies with darker pigmented flesh in the United States have a higher chance of being incarcerated, dying before the age of twenty-five, developing chronic illness, and earning less income.[12] These lived realities cannot be tied linearly or discretely to independent variables like education, class, or race. Black people in the United States do not die and get sick at higher rates than whites because they are Black; *they do so because of racism*.[13] The unequal distribution of wealth, education, and health outcomes is directly connected to racist environments. Racism is the original sin that shapes the other factors, a topic I return to later.

*Follow-Up Studies*

Regardless of the results of the intervention, the next steps are to follow up with the children from both the US and UK trials. The UK trial is also analyzing the biosamples for any potential biomarkers or indicators of GDM, to better understand mechanisms of metabolic disorders. The trial received £5 million to support the data analysis. In the end, the US consortium did not fund any follow-up studies to examine health in the offspring. Two trial sites from the US consortium applied for money separately and are following up with the children in their respective trials. The SmartStart trial administrators received money from a different grant to follow up with the children but decided not to collect DNA samples from the children. Instead, they are only examining behavioral variables related to eating and will collect weight measurements. Meanwhile, the UK follow-up study will take DNA samples from children and will collect a series of survey data from the mothers on eating and sleep patterns.

The SmartStart follow-up study with the children differs significantly from the UK study because the main focus of SmartStart is feeding. The study videorecords feeding encounters between the mothers and the toddlers in the homes of the participants. Grandparents, fathers, partners, and other caretakers are not a part of the study. In other words, the assumption of the feeding study is that the pregnant person becomes a mother, and the mother is the main individual responsible for feeding the child. This does not include partners, surrogates, or parents who go through the adoption process. It also does not include families who have extended family members as caretakers or paid caretakers who may also do the majority of the daily feeding. "The mother," and her implied role as the main nurturer, is imagined to be a woman, who may or may not work full-time.[14] The same gendered assumptions about maternal care and families are encompassed in the new designs. The imagined problems and solutions are reproduced.

In reflecting on these results, the US PI's conceptualization of behavioral interventions that target weight shifted in some significant ways, yet also remained unaltered. For instance, she explained that one of the key lessons of the trial was that "you get exposures postpartum, the *environment continues*—that is the take home." This part of her response reflects a greater awareness of environmental exposure beyond the critical period of gestation, on which the intervention was solely focused. However, the rest of her response reflects a similar solution to this new awareness of exposure. She went on to state, "You need to continue the intervention after pregnancy."

The inconclusive results shifted the focus toward a longer, more expanded period of time to apply the intervention, not necessarily to change the intervention itself. From the lens of postgenomic reproductive politics, preconception trials are an expansion of reproductive surveillance across the life course. While conceptions of exposure and environments shift, the individual lifestyle interventions—framed by a focus on weight as a key measurement and outcome—remain the same. This is epigenetic foreclosure: the simultaneous awareness that environmental exposures are complex, yet the intervention to address the multidimensional exposure over time remains the same.

## Politics of Postgenomic Reproduction: Racism, Reproduction, and Transgenerational Inheritance

Transgenerational inheritance characterizes the phenomenon of inheriting epigenetic modifications from past generations, in unpredictable and latent ways. Environmental exposure during critical periods is also used to explain transgenerational inheritance. The concept is used in ambiguous ways to help explain health patterns among children and adults who have been exposed to adverse conditions in critical periods of their lives, or in critical periods of their parents' and grandparents' lives.

The notion of "inherited trauma" may be used in uncritical ways to relate race and reproduction—for instance, to justify the high maternal and infant mortality rates among Black communities. Recent studies also draw from epigenetic theories to claim that brain plasticity is shaped by race—as an inherent attribute—instead of examining how racist environments shape brain development.[15] A foreclosed interpretation of transgenerational inheritance is the assumption that Black and Indigenous communities might already be predisposed to adverse health outcomes, regardless of medical interventions. These types of interpretations deny the fact that epigenetic modifications are malleable, unpredictable, nonlinear, and potentially reversible. We do not know when or how epigenetic modifications manifest and impact human health across the life course and across generations in human models. Epigenetics is connected to transgenerational trauma, and we can find instances of its manifestation, but it is not causal or predictable. Another issue with emergent integration of "race" into new epigenetics research is that it primarily focuses on racial difference, not on the impact of racist environments on health. Foreclosed translations of epigenetics obscure the reality that institutional racism, medical racism, obstetric racism, and daily encounters with anti-Blackness are directly related to the high rates of premature birth and death among Black, Brown, and Indigenous populations.[16]

Yet epigenetic foreclosure is not inevitable. There are alternative interpretations of epigenetics that explore how different kinds of environmental interventions can change health outcomes. For instance, Rachel Yehuda explores how stress and trauma can have biological impacts across generations. She characterizes the hope of epigenetics by stating that "just because you're born with a certain set of genes, you're not in a biological prison as a result of those genes—changes can be made to how those genes function."[17] One aspect of her work focuses on the children and grandchildren of Holocaust survivors; she found that the children of survivors had much higher rates of post-traumatic stress disorder. Although one negative and determinist interpretation of this finding might be that stress biologically and inevitably endures in bodies, Yehuda insists on a more optimistic interpretation. In an interview she did for the podcast *On Being with Krista Tippet*, Yehuda explained: "I'm very challenged by thinking how this information can be empowering and not disempowering. And one of the studies that we published [. . .] showed that some epigenetic changes occur in response to psychotherapy. If we're saying that environmental circumstances can create one kind of change, a different environmental circumstance creates another kind of change, that's very empowering."

The empowering part of epigenetics is that with environmental exposures, different physiological impacts can result. That is, if we can change environments, there may be biological impacts that follow. Yehuda is careful in attending to how epigenetics can "empower instead of disempower" because epigenetics can be interpreted in multiple ways. It is expansive, pluripotent, and plastic. In mainstream interpretations, the notion of transgenerational trauma provides a biosocial concept to understand pain and trauma in complex ways that are not necessarily determinable but also not invisible.[18] An alternative framework might examine how realities of land dispossession, racist health disparities, and death among Indigenous and African American communities are perpetuated not through epigenetic mechanistic pathways, but rather through long-term exposure to racist environments.

The disappointment with current applications of social and environmental epigenetics is its reductive interpretations and translations. For instance, as Yehuda claims, environments can change biological outcomes, yet in the clinical translation of epigenetics, the environment is framed as a discrete variable at the individual level, and not a complex multiscalar and temporal event. Most interpretations of epigenetics in prenatal trials narrowly constrain the environment to pregnant bodies and make those individuals responsible for changing their behaviors as if they were in complete control of epigenetic modifications. The same approaches are still enveloped in the more recent preconception trials.[19] The capaciousness of what can count as "environment" in epigenetics maintains political and reproductive stakes. These stakes center around questions like: Who and what are part of the maternal environment? And how are we held accountable?

*Foreclosed Alternatives*

During my concluding visit to the US trial, I asked the US PI: "If the trials are inconclusive, will you be doing different kinds of trials? Ones that do not focus on individual behavioral variables but perhaps other kinds of environmental factors?"

She said that she had submitted a grant to examine pesticides in the urine of the pregnant participants in the SmartStart trial. Since learning more about the population that participated in the trial, a population of primarily immigrant farmworkers from Mexico and Central America, she had become aware that the pregnant participants were exposed to high levels of potentially toxic chemicals in their work growing and harvesting produce in the United States. She conducted a preliminary study on the urine samples that the trial had collected, and the findings showed that participants in the intervention group had lower amounts of endocrine disrupters that are linked to pesticides and BPA. The interpretation was that because the trial had supplemented the participants' diet with meal replacements that were organic and BPA free, this nutritional component impacted hormone levels in the urine samples. She had begun the preliminary analyses on the urine samples and needed funds to do a more comprehensive analysis comparing the intervention and control groups to see if changing their diets had any impact on their pesticide exposure and levels of endocrine disrupting hormones. She applied for a small grant, a fraction of the amount that funded the SmartStart trial. Yet the conceptual shift in exploring other environmental factors was significant, an area the PI had yet to examine.

However, this new investigation was shut down. The grant was denied. This too is foreclosure: political and economic forces shape what kinds of epigenetic science is imaginable and possible. Of course, science is not produced in a vacuum, and it is not a neutral, objective process. We know from well-founded literature in science and technology studies that science shapes society and society shapes science, in a dialogical relationship.[20] Yet the point I want to emphasize is that sometimes "the social" is a stronger force in shaping what science gets done, what questions get asked, and whose bodies are targeted for interventions. In our current late liberal, capitalist, future-obsessed context, the maternal environment is still defined as an individual cis-body that is valued as reproductively capable. Supporting the exploration of different environmental variables in relation to health requires a shift in social values.[21]

Instead of looking at how toxic chemicals impact reproduction, the main grant that the PI received was similar to the SmartStart trial. The new trial was aimed at diabetes prevention, not obesity prevention, but applied the same lifestyle intervention as the SmartStart trial. The main difference and challenge in implementing the new trial was that the participants needed to receive the intervention *before* they became pregnant. This was a preconception, or prepregnancy, lifestyle intervention.

In my interviews with the staff who were working on this new trial, and who had worked on the SmartStart trial, they reported that the recruitment and intervention delivery were much more challenging. Fifty percent of pregnancies are unplanned. Targeting people before they are pregnant is a gamble. At first the staff focused on recruiting participants who had completed the SmartStart trial and said they wanted to have another pregnancy. The current trial implements the same intervention; once the participant gets pregnant, the intervention stops. So far, some participants have been receiving the prepregnancy intervention for about two years. No one has gotten pregnant yet; everyone is still waiting.

In the PI's own words: "Prepregnancy stuff is hot now, maternal weight before pregnancy is the question now." The trend of prepregnancy trials explains why it received generous funding, while the pesticide exposure during pregnancy study did not. Although this new trial is framed as "cutting-edge" because it is one of the first preconception trials, the conceptual design and focus on lifestyle behavioral intervention are not new. The preconception health campaigns that are the center of public health attention have now manifested in evidence-based clinical trials.[22]

These preconception trials also belong to future-oriented health logics. The hegemonic approach to addressing multidimensional chronic illness like diabetes and obesity remains the same. Now, not only are pregnant bodies risky for future health, but *any* body deemed reproductively capable is a potential risk and target for lifestyle interventions. The shift toward prepregnancy justifies the expansion of reproductive surveillance before, during, and after pregnancy. Surveillance expands, individual lifestyle interventions persist, and individual bodies remain the main environmental target. The heteronormative assessment of reproductive ability, long-term surveillance across the life span, and the inclusion of partner surveillance exemplify the politics of postgenomic reproduction of current and upcoming prepregnancy trials.

*Present Tense: Health Justice Now*

Exploring the findings of the US and UK trials and the ambiguous interpretations and applications of epigenetics is at once a narrative of foreclosure and potential. This journey in tracing the unfolding of epigenetics through pregnancy trials in twenty-first-century Western capitalist societies is unstable. It is our social values and imagined futures that must change. My own remaining questions have to do with how we might conceive of health justice in postgenomic reproduction. The core questions that orient my understanding of health justice are: What kinds of environments help us heal? What kinds of environmental changes can ameliorate adverse health outcomes? And how can we use evidence-based medicine like clinical trials in the service of health justice? Instead of focusing on how to change individual bodies to make them better fit and able to survive or be "resilient" in toxic, racist environments, we should ask, what kinds of environments promote healthy beings and relations?[23]

One area for exploring these questions is racism. Instead of examining race as a comparative category or racial differences, studies with a health justice focus might examine exposures to racist environments. I find that inconclusive comparative analysis across racial/ethnic groups can have the unintended consequence of obscuring processes of racism. How could this be? Existing approaches to racial disparities or disparities in health often make clumsy associations between ethnic categories like Black, white, and other with health outcomes like obesity and diabetes. In current framings of health disparities, Black and Brown people are assumed to have higher rates of obesity because they are Black or Brown. The solution is then narrowed down to changing individual Black and Brown people's bodies and behaviors. Yet the pregnancy trials at hand did not find any difference across racial/ethnic groups. As I described earlier, the explanation for this phenomenon is class and education. Finding no differences across race is not necessarily liberating or just; finding racism can be. The point is that disparities in health across race are not resolved by only including more ethnically diverse people and comparing their outcomes; such a narrow approach can actually obscure the impact of racism in health. By exposing these tensions and contradictions, the solutions that become possible might better address the core fissures and gaps.

Alternatively, applying the concept of racist environments opens up space to reframe problems and imagine alternative solutions.[24] By using the framework of racist environments, the problem changes: Brown and Black people have higher rates of obesity and diabetes due to long-term exposure to racist environments. Reframing the problem in this way promotes a different solution: What kinds of anti-racist interventions can ameliorate obesity and diabetes among Black and Brown groups? A justice-oriented approach to engaging with race in reproduction across science, medicine, and technology is to recognize and take seriously the processes of racism. Instead of comparing across races, studies applying an intersectional framework might focus on how processes of racism impact health and how these processes characterize our social and material environment.

New and old science already supports the finding that social inequalities and racism affect our health. Investing resources in studies that focus on individual resilience and its epigenetic consequences across generations is problematic. Resilience as a concept promotes an individualistic, essentialist notion that something inherent in one person can help them overcome enduring historical, systematic trauma; it obfuscates the responsibility and accountability that lie with institutions and systems. In practice, resilience research aligns well with economic and political efforts to dismantle social welfare programs and instead promote individual behaviors that can fit into existing violent structures. We know that racist, patriarchal, colonial, and capitalist environments make us sick. Making individuals more resilient in these environments will not change the fact that the same vulnerable people of color will continue to get sick and die at higher rates than others. Health justice now means that we ask questions about environments that are not individualized and questions about racism that are not about racial/ethnic comparisons. Overall, refocusing attention and resources through frameworks of justice and equity, and studying the health impacts of racism directly can illuminate alternative solutions.

Finally, justice-oriented health in reproduction focuses on our collective entanglements in birthing and rearing children who depend on healthy environments to sustain life, human and nonhuman alike.[25] I began this book with the provocation that we all collaboratively engage in reproductive processes, regardless of sex, gender, and capacity. Reframing what counts as the maternal environment to include collective practices of reproduction beyond cis-gendered, reproductively-capable bodies expands the scope of solutions: What kinds of parental leave options are available to biological/nonbiological siblings, cousins, friends, and neighbors? What forms of maternal health policies serve and protect transgender communities? Feminist and critical race frameworks are vital in the reexamination of racialized, heteronormative, and individualistic values inherent in existing interpretations of new science, approaches to child maternal health, and evidence-based medicine. Applying these frameworks in the production of scientific knowledge reconceptualizes current health problems and opens up alternative solutions.

# Epilogue

[THE FUTURE] IS COMPOSED OF NOWS

*Summer 2020*

The virus confines movement and contact.

Air travel is deemed unsafe, so I drive twenty-one hours from Massachusetts to Florida to support my pregnant sister with childcare. On my drive, I stop in Washington, D.C., where my younger sister lives, 1.3 miles from the White House. For most of the seven-hour drive down to D.C., I listen to the news coverage on the growing protests and marches. More murders: Ahmaud Arbery, Breonna Taylor, and George Floyd.[1]

They said it was a "tipping point."

Four months of quarantining, the highest unemployment rates in American history, the stock market crash, the disproportionate deaths and impact of COVID-19 on Black and Brown communities, and another viral video of a Black American being killed at the hands of the state.

My sister and I joined the protests around all those white buildings and monuments. We wore masks. I heard the words, "No one is coming to save us."

*Fall 2020*

Rising rates of COVID-19.
    Elections, virtual debates, long lines to vote.
    Effective vaccine emerges.
    Election results: Biden and Harris win.
    Results contested.
    The one-term president kept saying "fraud, stolen, fraud, stolen."

*Winter 2021*

I am driving back up from Florida to Massachusetts.
    My niece Mila was born. Her parents wore masks at her birth. We could not go to the hospital.
    I'm getting closer and closer to D.C. Text: "Have you heard the news?"
    Text: "Where are you staying in D.C?"
    Text: "They stormed the capital!"
    I turn on the radio and hear the words "domestic terrorism" and "insurrection!" I turn off the radio. I make it to Alexandria just before the mandatory curfew.

*January 20, 2021*

Biden and Harris are sworn in.
    Slowly, chaotically, the vaccine is distributed.
    Reprieve?

.    .    .    .    .

The vignette above illustrates part of the context that shaped the writing of this book. While the implications of these events will be analyzed for many years to come, here I gesture at the relevant stakes of this book within the current moment. I will briefly interrogate our public understanding of clinical trials for vaccination development and how diversity and inclusion is currently imagined as distinct from equity.

As of July 2021, more than 600,000 people had died of COVID-19 in the United States, over 100,000 deaths had been recorded in the United Kingdom, and there had been over 4 million deaths globally.[2] The political and public health crisis permeated each moment of 2020 and still dominates our everyday lives. Black and Brown people continue to die at disproportionate rates. During the development of the vaccine, some media coverage (mostly on National Public Radio) reflected on the issues of recruiting diverse people into clinical trials. Once the vaccines were approved, some small news coverage emerged on the disparities in distribution and access.

Who we imagine participating in clinical trials is often misrepresented and misunderstood. That is, the blanket statement that there is not enough racial diversity in clinical trials is both accurate and inaccurate; while there aren't enough "racial minorities" in phase II or III clinical trials, African American and Latinx populations may be *overrepresented* in phase I clinical trials, the riskiest phase of clinical trials.[3] In the book *Adverse Events: Race, Inequality, and the Testing of New Pharmaceuticals*, Jill Fisher writes that contrary to the public imaginary, "Men and racial minorities are the predominant phase I trial participants."[4] Fisher's research on phase I clinical trials in the United States found that financial incentives were one of the key forms of recruitment and retention, and that "Blacks and Hispanics make up a disproportionate number of healthy volunteers" in phase I trials.[5]

In part, the misconceptions about representation occur because there are different types, kinds, and phases of clinical trials, and we do not have enough national data on trial participation across a large variety of trials. For instance, there can be up to four phases in a clinical trial, there are public or private trials, there are different target populations (pregnant or nonpregnant adults), and there are different types of trials for pharmaceutical development or a behavioral intervention like the ones that I examined in this book. Together these factors can paint a very different picture of who is included and represented in clinical trials.

While it is true that for the development of the COVID-19 vaccine it was necessary to include diversity, what diversity actually means for pharmaceutical development relates to body size, age, environmental exposure, and genetic variation that may impact drug metabolism and response. Ethnoracial classifications of "race" as an index of diversity can be misleading. Dr. Alexa Volpe, a clinical pharmacist who is assisting in the vaccine distribution in San Antonio, Texas, commented at a lecture in my "Race, Gender, and Science" seminar: "'Race' is a poor proxy for genetic polymorphisms. For example, many Latinx individuals can also use the classification 'white' for race to identify, however, there is great genetic diversity within the Latinx population." Volpe echoes the well-established understanding that there is more variation within imagined race/ethnicity groups than between such groups.[6] The important distinction to make is that Black and Brown people are needed in clinical trial development not because they are Black or Brown in skin pigmentation or ethnic and racial identification. Black and Brown bodies do not metabolize drugs differently because they identify as a certain race or ethnicity. The narratives around diversity often tread a narrow line between essentialism and deterministic biology on the one hand, and on the other liberal democratic notions of inclusion. All different kinds of bodies are needed in clinical trial research. And although racial classification of difference is not biologically based but rather socially, historically, and politically constructed, processes of racism can physiologically impact bodies differently. Epigenetics provides a biosocial mechanism for understanding how experiences of racism impact health.[7] Yet studies that draw on epigenetics continue to use racial classification of difference as a key framework for understanding diversity, not processes of racism or issues of equity. The point of highlighting this issue in the unfolding of COVID-19 is similar to the argument I make in this book: focusing too much on diversity and inclusion via proxies of race/ethnicity classifications often distracts from effective or precise characterizations of "diversity" needed in science that directly address racist disparities in health and not just racial difference.

The "cruel optimism" of the diversity politics inherent in the unfolding of COVID-19 is that it provides fundamental lessons on how to, as Samuel Beckett wrote in 1983, "Fail again. Fail better."[8] The desire for diversity and inclusion, narrowly defined as including more Black and Brown people, inhibits a more generative, precise, and effective approach to addressing health disparities. There is optimism, however, in the awareness of this failure. Recognizing the current limits of diversity may propel a different approach, which might also fail, but fail slightly better. In this way, foreclosure can be generative, failure can be generative. For instance, an alternative approach to the problem of health disparities in obesity and diabetes across race would be to reframe the problem as prolonged exposure to racism, not an individual failure of will and discipline. Doing so recognizes how racist environments impact chronic disease and enables the creation of interventions in contexts and structures and not solely among individuals. Likewise, reframing the problem of diversity in clinical trials as more than just a lack of different ethnic or racial participants, but rather as a problem of systems of power, can also address issues of equity in scientific knowledge production.

At the edges of foreclosure and failure are possible ways of doing and being otherwise. Investing in the interruption of systemic, infrastructural, and environmental racism can have lasting and meaningful consequences for medical and scientific innovation and creativity.

# Notes

INTRODUCTION

1. See Michelle Murphy's (2011) concept of distributed reproduction; Hannah Landecker's (2011) discussion of lateral reproduction; see also Briggs (2017); Ross and Solinger (2017); Lewis (2019).

2. Postgenomics represents a significant shift in approaches and scientific content, including "research on gene-expression, population-level genetic variation, and gene-environment interactions" or (GEI) research (Ackerman et al. 2016, 215). See also Richardson and Stevens (2015) for expanded definitions of postgenomics.

3. Reardon (2017).

4. Briggs (2014); Wartman (2013).

5. Richardson and Stevens (2015); Richardson et al. (2014).

6. Mansfield and Guthman (2015).

7. Sullivan (2013).

8. Frew et al. (2014). I use the term *pregnant people* because not all women reproduce, and not all pregnant bodies identify as cis-women. I also use the term *woman*, which was the term used in my fieldwork.

9. National Institutes of Health (2014).

10. Not all prenatal trials draw from DOHaD and epigenetic science; although many DOHaD and epigenetic studies on human models focus on pregnancy and reproduction, there is an expansive area of epigenetic research on plants, insects, and other animals.

11. Fisher (2009).

12. Epstein (2009); Fisher and Kalbaugh (2011).

13. Currently, all states provide prenatal care and up to sixty days of postpartum care to anyone who is low income and does not have health care. However, the services provided by Medicaid and private insurance vary widely across the country and within states.

14. These services include ambulatory and emergency services, as well as radiation, chemotherapy, and complex surgery.

15. Holmes (2021).

16. Central Intelligence Agency (2015).

17. Davis (2019).

18. Some estimates reflect that Black British women are five times more likely to die during childbirth or experience pregnancy complications (Knight et al. 2018).

19. Murphy (2017a).

20. Foucault (1994b).

21. Although there is no consensus on how to use terms like *neoliberalism* or *late liberalism*, I employ the terms *late* and *contemporary liberalism* as a way to index particular aspects of the current political and economic climate. For instance, while late liberalism is deeply connected to neoliberalism in that both encompass an enduring focus on individuals taking care of themselves in the context of a retreating welfare state (see the work of Fredrick Hayek [1994] and Michael Polanyi [1975] on neoliberal theories from the 1940s and 1950s, which reached state policies in the 1970s in the United States and United Kingdom), late liberalism is also connected to the expanded neoliberal policies of the 1980s and 1990s, which dismantled the social safety nets in the United States and United Kingdom. Such policies promoted international free trade, limited unions, and deregulated corporations (Brown 2019; Harvey 2007). Contemporary or late liberalism spotlights the cumulative impact of neoliberal policies from the past half century (Povinelli 2016).

22. Scholars call this pressure to make sure one's body is "healthy" for the sake of national belonging *biocitizenship* (Greenhalgh and Carney 2014; Molina 2006; Rose 2006).

23. Landecker and Panofsky (2013); Mansfield and Guthman (2015); Meloni (2017).

24. The focus on systems and relations of power is important to the analysis because individual scientists do not decide on funding trends or the scientific methods that are supported by medical infrastructure. In addition, in the current economic and political climate, it is impossible for every individual to achieve the social and health mobility that is required to be a "good" citizen: coded as thin, white, middle class, nondisabled, and hetero/homonormative.

25. KHP conference (2014; emphasis added).

26. To be sure, there are many different approaches and agendas within maternal health and health policy. The NHS is currently working on a multiyear prevention initiative and is actually actively giving money to health-care projects that can address prevention/prediction. My aim is to spotlight how influential these epistemic environments are in shaping what science gets done and what interventions get funded.

27. Rose (2006); Rajan (2006).

28. Anticipatory regimes are defined through capitalist ideologies, described in the next sentence, as individual self-care, interventions aimed at prevention of risk, and the anticipation of disease (Adams, Murphy, and Clarke 2009). For characterizations of capitalist ideologies shaping health see Cooper (2008); Dumit (2012).

29. El Haj (2007).

30. Adams, Murphy, and Clarke (2009).

31. Valdez and Deomampo (2019). For instance, mass incarceration directly impacts the reproductive lives and futures of Black families through disproportionate surveillance, unstable housing, and unemployment (Roberts 2009; Gilmore 2007).

32. Sasser (2014); Hoover (2017, 2018).

33. Even though my analysis here focuses on capitalism—it is important to highlight that neoliberalism, and settler late liberalism, as Povinelli (2016) writes —are directly entangled with late capitalism.

34. Baptist (2016); Desmond (2019)

35. I ground both surveillance and biocapitalism in racial capitalism because these relations have predominately been obscured. Despite the substantial work of Cedric Robinson (2000) and the Black Marxist tradition, discussions surrounding surveillance and biocapitalisms obfuscate the long-standing connection between racism and Western capitalism. One example of this is in Lock's examination of the commodification of bodies in "black markets," which does not recognize the role of slavery or racial capitalism in the alienation of bodies (Farquhar and Lock 2007). A more recent example, is Zuboff's (2019) text on surveillance capitalism. Her text does not reference slavery as a foundational form of theft and dispossession that shapes capitalism. An example of a text that mediates this gap in the literature is Weinbaum's (2019) work, which draws heavily on Black feminism to connect biocapitalism to racial capitalism through a focus on reproduction.

36. See the work of Hartman (2008); Spillers (1987); Browne (2015); see also Sylvia Wynter's collected work (2003) and McKittrick's analyses of Wynter (2006, 2021). Transnational and postcolonial feminism also contributes to theorizing the power relations of race, gender, labor, and migration in relation to capitalism (Mohanty 2003).

37. Weinbaum (2019).

38. Harvey (2005). See also Arendt (1973) and Luxemburg (1913); Rajan (2012).

39. Stout (2019); Appel (2019).

40. Shoshanna Zuboff (2019) calls this *behavioral future markets*.

41. Zuboff (2019).

42. Rosato (2020)

43. Rosato (2020).

44. Rajan (2006).

45. See Cooper's (2008) work on bioeconomies; Waldby and Mitchell's (2006) work explains the political economy of tissues and the notion of biovalue; together Cooper and Waldby (2014) expanded a Marxist feminist analysis of labor and value in relation to donors and research subjects in a global bioeconomy. See also Vora (2015) for feminist labor analysis of biocapitalism.

46. For racialization and commodification of gametes see Deomampo (2019); for eugenic perspectives of egg selection see Martin (2018); for a global perspective of reproductive waste see Kroløkke (2018).

47. See Risa Cromer's (2018, 2019) work on how the ambiguous use of embryos unfolds in scientific research and Christian adoption agencies.

48. See Dumit (2012) and Peterson (2014) for drug markets and speculation. See Weinbaum (2019) regarding assisted reproductive technologies; see also Atanasoski and Vora (2019) discussing racial capitalism and surrogacy labor. See Cooper and Waldby (2014) and Landecker's chapter on HeLa cell lines (2010).

49. Although the US Supreme Court changed patent laws in relation to biomarkers, stating that "patents cannot cover discoveries of basic correlations in nature, such as those relating biomarkers to particular clinical outcomes," the process of integrating biomarkers into clinical care can still "earn intellectual property protection" (Kesselheim and Shiu 2014, 127).

50. For instance, in an article from the NHS, the authors explained that predicting diabetes in the future is so valuable that the potential cure can justify the disinvestment in public health care in the present (Farrar et al. 2016).

51. To be clear, precision medicine is not to be conflated with predictive approaches. For instance, particular genetic biomarkers and drugs like Tagrisso are effective forms of precision medicine. While we do not need predictive biomarkers to tell us who will get sick, this does not discount the value of precision medicine entirely.

52. Martin (2007).

53. While postgenomics is often referred to as a paradigm, this is still contested (Meloni and Testa 2014). See also Meloni's (2018) review at Somatosphere of Reardon's book the *Postgenomic Condition*. Regarding genes and human variation and development, see Barnes and Dupré (2008).

54. My characterization of postgenomic reproduction is a working definition, by no means all encompassing.

55. The study of cloning and research in assisted reproductive technologies (ARTs) like IVF were framed as *reprogenetics*, a term used in the late 1990s (McKee et al. 2003). For research examples in the area of ARTs, see Charis Thompson (2007), Sarah Franklin (2013), Marcia Inhorn and Birenbaum-Carmeli (2008). For specific work on surrogacy and race see Daisy Deomampo (2016). For the intersection of embryo adoption/donation, race, and religion see Risa Cromer (2018, 2019, 2020). For past and present work across reprogenetics and epigenetics see Rayna Rapp (2018), Elizabeth Roberts (2017), and Janelle Lamoreaux (2016).

56. See Ginsburg and Rapp (1991, 1995).

57. For instance, the notion of stratified reproduction describes the ways in which economic and social forces create the conditions under which certain people's reproduction is valued, while the reproduction of others is denigrated. See Shellee Colen's (1995) chapter in the volume edited by Ginsburg and Rapp (1995), also referenced in Valdez and Deomampo (2019).

58. The RJ framework prioritizes the stories of women of color as the foundation for new knowledge (Ross and Solinger 2017; Silliman et al. 2004).

59. See Briggs (2017)2.

60. Price (2010). Megan Warin and I wrote about how discourses of choice were mobilized by political agendas during the COVID-19 pandemic, which masked the underlying racist reality that not all people are afforded the same rights or options from which to choose. The underlying message of slogans like "my body, my choice" is that "*white* bodies have choices" (Warin and Valdez 2020).

61. Strategies of bodily surveillance, interventions, and management are indicative of the biopolitics of governmentality (Foucault 1978; Cohen 2009; Thacker 2005; Clough and Willse 2011).

62. Foucault (2003).

63. The book *Imperialism and Motherhood* offers a feminist analysis on how women's bodies and behaviors were the main targets of national health interventions (Davin 1978).

64. See Rafter (1992) and Tomes (1999) for examples of public health campaigns from the nineteenth and twentieth centuries that targeted poor women to reduce their fertility rates and prescribed a "domestic science" of house cleaning for germ prevention. See Graham (1996) and Wolf (2003) for examples of public health interventions focused on alcohol and tobacco consumption and breastfeeding.

65. Almeling (2020).

66. Rapp (1999); Ross and Solinger (2017).

67. Ross and Solinger (2017); Roberts (1998); Davis (2019); Nash (2021).

68. Roberts (2009); Bridges (2011); Edu (2018); Murphy (2017b); Clarke and Haraway (2018); Saldaña-Tejeda (2017); Nash (2021). See also the work of the

Race and Reproductivities working group, which includes, Davis (2019), Falu (2019), Cromer (2019), Deomampo (2019), Valdez (2019b) and Craven (2019).

69. Lopez (2008); Briggs (2017).

70. Lappé, Hein, and Landecker (2019).

71. Lappé, Hein, and Landecker (2019).

72. See Rose (2006) for an explanation on the optimization of life.

73. Waggoner (2017). See also Pentecost and Meloni (2020).

74. Almeling (2020).

75. For work on reproduction in a postgenomic era see, Sarah Richardson's forthcoming book and the forthcoming special issue of *Science, Technology and Human Values* on postgenomic reproduction, guest edited by Jaya Keaney and Sonja van Wichelen. For emerging work on CRISPR technology, see Santiago Molina's (forthcoming) dissertation, "How Science Produces Institutions: The Practice and Politics of Genome Editing."

76. Increasingly, epigenetic studies on human populations are not finding the expected modifications and impacts across maternal and infant health outcomes. See Zeneta Thayer and Amy Non (2015).

77. Thompson (2007).

78. See, for instance, Dupras and Ravitsky (2016).

79. I am grateful to Banu Subramaniam for her inspiration in framing the geneticization of life.

80. For conceptualizations of situated knowledges see Haraway (1988) and Haraway (1990b), which draws on Nigerian author Buchi Emecheta.

81. I write *race/racism* to recognize how notions of race are always already imbricated with forms of racism. The inception of race as a classificatory system was based on an envelopment of racism in science since the Enlightenment (Duster 2006, 2005; El-Haj 2007).

82. For instance, Lorraine Hansberry, James Baldwin, Nina Simone, Audre Lorde, Claudia Rankine, and Kara Walker.

83. I use *queer* as a way to describe my way of thinking and being in the world, which is not fixed into any single or stable manifestation, aesthetic, desire, or attraction. It shapes my approach to theory, analysis, practice, and relations. I use the concept of queerness with the recognition that it is rooted in imperialist legacies of the English language.

84. Anzaldúa (2012).

85. Most prominently, the canons of Black feminism and Black studies have built a rich archive of evidence through experiential testimony, law, and history that undeniably establishes the existence and processes of racism, racialization, and institutional/systemic racism. Ruth Gilmore defines racism as "the state-sanctioned and/or extralegal production and exploitation of group-differentiated vulnerability to premature death" (2007, 247). Racialization is a process that links

racialized qualities to people, groups, or things; such qualities or categories are constantly evolving (Banton 1977; Miles 1989; Omi and Winant 2015). See also the definition of race/racism from the American Association of Physical Anthropologists at https://physanth.org/about/position-statements/aapa-statement-race-and-racism-2019/.

86. Wynter (2003); McKittrick (2015).

87. "Racializing biological matter (M'charek 2013), bodies (Haraway 1989; Schiebinger 1993), or flesh (Wynter 2003) contributes to social divisions across 'humans' and particularly, those who have never been deemed fully human (see Wynter via Weheliye 2014)"; Valdez and Deomampo (2019, 552).

88. My thinking is shaped by notions of race craft, race as technology, racialization, obstetric racism, fugitivity, race as material-semiotic and relational, racism as alienation, degrees and genres of humanness, racial classification as power of naming and claiming, misogynoir, and white empiricism. See the work of Chen (2012), M'Charek (2013), Omi and Winant (2015), Benjamin (2019), Haraway (1997), Davis (2019), Edu (2019), Harney and Motten (2013), Wynter (2003), Spillers (1987), Bailey (2021), Prescott-Weinstein (2020), Subramaniam (2014), Nelson (2016), El Haj (2007), and Agard-Jones (2013).

89. See McKittrick's *Demonic Grounds* (2006), in which she writes that "racism and sexism are not simply bodily or identity based; racism and sexism are also spatial acts" (xviii).

90. Davis (2019); Hoberman (2012); Benjamin (2019). Comorbidities of obesity and diabetes are just one contributing factor to the higher death tolls among Black and Brown communities during the COVID-19 pandemic (CDC 2021).

91. Slopen et al. (2015).

92. Natali Valdez (2019a).

93. Hartman (2019).

94. See McKittrick's (2021, 153) recent book for examples on how to create new methodologies that "breach" antiblackness.

95. Various small parts of these chapters appears in *Medical Anthropology* (Valdez 2019b), *American Anthropologist* (Valdez 2020) and *Science, Technology, & Human Values* (Valdez forthcoming).

## CHAPTER 1. EPISTEMIC ENVIRONMENTS

1. Shiell et al. (2001).

2. Shiell et al. (2001).

3. Grieve (1974,1975); Grieve et al. (1978).

4. University of Edinburgh (2005).

5. Shiell et al. (2001, 1283).

6. Shiell et al. (2001); Herrick et al. (2003); Reynolds et al. (2007).

7. Hsu and Tain (2019).

8. See table 1.

9. Barnes and Dupré (2008).

10. Barnes and Dupré (2008).

11. For other examples of epigenetics and its sociological significance, see Hannah Landecker (2011), Maurizio Meloni (2016), Becky Mansfield and Julie Guthman (2015), Margaret Lock (2013), and Jörg Niewöhner (2011).

12. Genome wide studies (GWAS), and epigenome wide studies (EWAS) represent the postgenomic technology that examines the sequencing of genomes and epigenomics, the latter of which includes not just the genes but the material on and around genes that modify how the parts of the DNA are transcribed to make proteins. See also Richardson and Stevens's (2015) definition.

13. Szyf (2009); Holliday (2006); Hurd (2010). In addition, epigenetics refers to genome-environment interactions and epigene-environment interactions (Lappé g and Landecker 2015).

14. Bird (2002); Feinburg (2007).

15. Epigenesis explains how an embryo gradually becomes more complex through repetitive cell division. In a letter to the journal *Nature*, Waddington explains the significance and derivation of the term epigenetics in relation to the debate on epigenesis and preformation: "The study of the 'preformed' character nowadays belongs to the discipline known as genetics; [. . .] Admittedly the word 'genetics', which was coined by Bateson to cover the 'the physiology of descent', might have been used so as to embrace both [biological and genetic] development; but in practice it has not been widely employed in [the genetic] sense" (1956, 1241). The emphasis that Waddington places on epigenesis and its interaction with "preformed characters" is the connection between genetics and developmental biology.

16. Szyf (2009); Holliday (2006); Hurd (2010).

17. Waddington (1956).

18. Lamarck published his findings in the early 1800s, which contributed to an understanding of "soft inheritance," or the inheritance of acquired characteristics. In late 1871 another theory of inheritance emerged and was developed by Alfred Wallace and Charles Darwin, which specified that changes in species occurred through the process of "descent with modification." Darwinian theories prevailed, and Lamacrkism dissolved as a legitimate theory.

19. Around the same time that Darwin was publishing, Gregor Mendel established his theory of genetic inheritance. Through pea plant models, Mendel developed our modern understanding of how certain genes are inherited from one generation to the next.

20. Meloni and Testa (2014).

21. Szyf (2009, 9).

22. Landecker and Panofsky (2013).

23. Weaver, Meaney, and Szyf (2006).

24. Van Speybroeck (2000); Shostak and Moinester (2015).

25. Darling et al. (2016, 58).

26. Darling et al. (2016).

27. Shostak (2013).

28. Lock (2013).

29. El-Haj (2007).

30. Feng, Jacobsen, and Reik (2010).

31. Niewöhner (2011, 284).

32. Mansfield and Guthman (2015).

33. Landecker and Panofsky (2013); Lappé and Landecker (2015).

34. Exact quote: "Maternal diet and nutrition status during pregnancy also *predict* risk for [cardiovascular disease, blood pressure, insulin resistance, type 2 diabetes, and a tendency to deposit fat in the central, metabolically active fat deposit]" in the children (Kuzawa and Quinn 2009, 133).

35. Wadhwa et al. (2009).

36. Barker and Osmond (1986).

37. Barker (2007).

38. Barker et al. (1993).

39. See Roseboom et al. (2001) for the Dutch study. See Bygren et al. (2014) for the Swedish cohort study.

40. Syddall et al. (2005).

41. Syddall et al. (2005, 1237; emphasis added).

42. Wadhwa et al. (2009); Barker (1995); Oken and Gillman (2003).

43. McGoey et al. (2011); Cartwright and Hardie (2012); Cohn et al. (2013).

44. Fisher (2009, 2020); Petryna (2009).

45. It is important to note that studies designed and funded by wealthy countries in the Global North follow distinct ethical guidelines when they are implemented in places like India, Brazil, and Africa (Murphy 2017a; Petryna 2009; Sunder-Rajan 2009; Yates-Doerr forthcoming). My depiction of recruiting is specific to the United States and United Kingdom.

46. Petryna (2009); Sunder-Rajan (2006); Palmer (2010); Pollock (2012).

47. Palmer (2010). Anderson (2006) argues further that the laboratories of hygienic practices were not only teaching and disciplining hygiene; they were also testing the effectiveness of subject formation.

48. Palmer (2010, 1).

49. The RCT method was not based solely in the clinic or in medical laboratories. In the 1920s, F. S. Chapin from the University of Minnesota and Ernest Greenwood at Columbia University used derivations of the current RCT method

to study topics related to rural education, public housing, and social interventions for "delinquent" boys (Oakley 1998). These older studies on delinquency are akin to current behavioral studies that use epigenetic ideas to explore "aggressive" behaviors in children through early interventions. See also the chapters by Stringhini et al., Wells et al., and Kelly et al. in Meloni et al. (2018). The laboratories in colonial settings expanded to global networks of institutes that fed into the London School of Hygiene and Tropical Medicine, the Antwerp Institute for Tropical Medicine, and the Royal Tropical Medicine in the Netherlands, all of which became part of a movement toward global epidemiology and eventually the establishment of the World Health Organization (Lock and Nguyen 2010).

50. One such infamous example is the Tuskegee study in the United States, which unethically examined the development of syphilis in Black Americans, despite the knowledge of an emergent cure (Reverby 2013). Another example is the unethical experimentation on Navajo uranium miners in the 1950s, despite the known link between uranium exposure and cancer (Brugge and Goble 2002).

51. Thalidomide was marketed to pregnant women as an antinausea medication, primarily in Europe. This drug was only tested on male rats, and its effect on pregnant people remained unknown (Saunders and Saunders 1990).

52. Rehman, Arfons, and Lazarus (2011).

53. Parisi et al. (2011).

54. Cooper (2011, 83).

55. Sturdy and Cooter (1998).

56. Oakley et al. (2004).

57. Oakley et al. (2004).

58. See Croneback and Snow (1976), cited in Oakley et al. (2004, 441).

59. See Bhaskar (1975), cited in Oakley et al. (2004, 442).

60. There were other large pregnancy trials that all implemented similar lifestyle interventions. See the LIMIT study, "Limiting Weight Gain in Overweight and Obese Women during Pregnancy to Improve Health Outcomes," one of the largest studies testing lifestyle interventions of diet and exercise on overweight pregnant women, in Australia in 2012 (Dodd et al. 2014). The study was inconclusive (Dodd et al. 2018). See also the ROLO study, in Dublin, Ireland, which also implemented a low glycemic index diet in pregnancy to prevent macrosomia (Walsh et al. 2012); it also was inconclusive.

61. All names of people and the trial itself have been changed.

62. US consortium (2018) (authors' names are not cited, to maintain anonymity).

63. In the United States there was a working group funded by the National Institute of Diabetes and Digestive and Kidney diseases (NIDDK) called Obesity Intervention Taxonomy and Pooled Analysis Working Group.

64. To be sure, there are significant differences, such as new statistical technologies and massive amounts of data collection procedures. Yet another enduring

aspect, which applies to the United Kingdom only, is that the structure that entangles health-care services with research exists today. Dr. Grieve was a health-care provider and a researcher. The National Health Service (NHS) still promotes the three-pronged approach of integrating teaching, research, and health care.

65. See the work of Londa Schiebinger (2004), Ian Hacking (2001), Charis Thompson (2007), Donna Haraway (1997), and Bruno Latour (1993). For notions of coproduction of science and society see Jasanoff (2007); Reardon (2004).

66. Pollock and Subramaniam (2016); Bailey and Peoples (2017).

67. Nelson (2016); Benjamin (2013, 2019); Noble (2018).

68. Haraway (1988, 1990a, 1997); TallBear (2013).

69. Allen and Perry (2012); Alcoff (2015).

70. Prescod-Weinstein (2020).

71. Prescod-Weinstein (2020, 421).

72. Schiebinger (2004).

73. Subramaniam (2014); Murphy (2017a); Strings (2019).

74. Schwartz (2001).

75. Currently, African American and Latinx communities are suffering and dying from COVID-19 at disproportionate rates. See CDC (2021). The surgeon general suggested that the high rates of COVID-19 among communities of color are related to risky behaviors like drinking or smoking too much (Sellers 2020). His comments obfuscated the reality that repetitive long-term exposure to racist environments directly shapes the unequal access to safe and affordable house, health care, nutrition, and employment.

76. El-Haj (2007).

77. Meloni (2016, 2017).

78. Meloni (2014).

79. NIH (2021).

80. Meloni (2015); Mansfield and Guthman (2015).

81. Shiell et al. (2001).

82. Shiell et al. (2001, 1283).

83. Abu-Lughod (1990); Stacey (1988).

84. Subramaniam (2014); Gordon (2008).

85. Specifically, Hancson works at the same university Barker did in Southampton, and they coauthored papers together.

86. This comes from the Amazon page that summarizes the book: www.amazon .com/Mismatch-Lifestyle-Diseases-Peter-Gluckman/dp/0199228388.

87. See the work by Nutrire CoLab (2020), whose members include Diana Burnett, Megan A. Carney, Lauren Carruth, Sarah Chard, Maggie Dickinson, Alyshia Gálvez, Hanna Garth, Jessica Hardin, Adele Hite, Heather Howard, Lenore Manderson, Emily Mendenhall, Abril Saldaña-Tejeda, Dana Simmons, Natali Valdez, Emily Vasquez, Megan Warin, and Emily Yates-Doerr.

88. Gluckman and Hanson (2008).

89. Gluckman et al. (2005, 530).

90. Though many pregnancy interventions and experiments have occurred and continue to take place in colonial and postcolonial contexts, the studies I focus on in this book are grounded in the Global North and are influential in shaping the fields of environmental epigenetics and Developmental Origins of Health and Disease (DOHaD). There are many examples and a long history of managing reproduction through prenatal interventions in places like West and Central Africa, Guatemala, the Philippines, and Bangladesh. See, for instance, Murphy (2017); Yates-Doerr (forthcoming).

91. Sharp, Lawlor, and Richardson (2018).

92. Almeling (2020).

93. Valdez (2018).

94. The rise of fetal personhood in the twentieth century is symptomatic of how pregnant persons are viewed in a heteropatriarchical society in general: pregnant bodies are assumed to be cis-gendered women, synonymous with mothers, and public property that is subject to state surveillance and control. All of this is also racialized through the stratification of reproduction (Colen 1995; Casper 1998; Duden 1993; Rapp 1999).

95. Deomampo (2016, 2019).

96. Clarke and Haraway (2018); Smietana, Thompson, and Twine (2018).

97. I posit that nascent shifts toward preconception trials are a direct response to the inconclusive results from pregnancy trials.

98. Richardson et al. (2014).

99. Hales and Barker (1992).

100. In order to preserve the anonymity of my informants, I do not cite or quote from published articles directly but instead paraphrase key messages.

101. Strathern (2001, 2004, 2020).

102. Marcus (1995); Marcus and Fischer (1999).

103. Abu-Lughod (2000); Rajan (2006).

104. See Hartman's (2019) book, which explores the wayward lives of the unseen Black young women in the nineteenth and early twentieth centuries.

105. For methodological discussions on improvisation, see Dumit (2018).

106. Another relevant question is: What counts as a comprehensive scientific paradigm shift? In *Structures of Scientific Revolution*, Kuhn examined historical records to argue that scientific discoveries are not isolated or distinct events credited to individuals but part of a larger process and development of knowledge production. Kuhn (1996, 11) defines paradigms as having two main characteristics: first, they must be unprecedented to attract people toward them and away from competing notions; and second, the achievement must also be open ended enough to allow the "profession" to solve problems. However, what politics and power dynamics must be maintained if reaching consensus means that

predominantly the same people from the same geopolitical regions get to contribute and define past, present, and future paradigms?

CHAPTER 2. UN/ALTERED

1. Siega-Riz et al. (2020).
2. Siega-Riz et al. (2020).
3. Scott et al. (2014).
4. NICE (2010).
5. Lampland and Star (2008).
6. Dumit and de Laet (2014).
7. Van Eijk, Marieke (2018).
8. Currently, the focus on calories instead of quality of food substance remains relevant. The US SmartStart intervention focuses on weight control via calorie control, not the kind of calories or food quality.
9. Today the etiology of preeclampsia is still unknown, but new research is showing that it may be triggered by exposure to the fetal genome that crosses the placenta as particulate matter (Kenny and Kell 2018).
10. Brewer and Brewer (1983).
11. Bell (2010); see also Paul (2010).
12. Almond and Currie (2011).
13. Susser and Stein (1994).
14. Weir (2006) shows how the development of epidemiological risk became entangled with notions of "perinatal mortality," a notion conceived in the 1950s and defined as the death of a fetus or newborn before or after birth. Together risk assessment and perinatal mortality provided legal and medical justification for state-sponsored interventions on pregnant women, mothers, or families. Thus, any form of risk that is framed as potentially harmful to the fetus can be legally and medically used to control pregnant women's bodies and behaviors. This gendered and political context lays the groundwork for understanding how and why pregnant women's bodies are available for experimentation, surveillance, and management in contemporary pregnancy trials.
15. The 1970 National Research Council (NRC) publication on maternal nutrition and weight also made it mandatory to provide prenatal vitamins (specifically folic acid) for pregnant women (NRC 1970).
16. See NRC (1970).
17. IOM (1985).
18. NRC (1970).
19. Another study that was also influential in the conversation of maternal nutrition and health was the US Collaborative Perinatal Project (CPP). The CPP study was implemented during the 1950s and 1960s, and it focused on ways to

improve pregnancy through different approaches (Rush 2001). The CPP study concluded that negative health outcomes during famine did not just have to do with weight, diets, or calorie intake; rather, the social conditions during pregnancy played a role in child maternal health. See Hardy (2003).

20. See forthcoming work by Emily Yates-Doerr.

21. Interview with AN, March 10, 2014. My qualitative data are referenced in this manner: type of encounter (interview or observation), date, and participant's initials.

22. Rush (2001).

23. For neoliberalism see Brown (2019) and Harvey (2007). For neoliberalism in relation to race and reproduction see Ross and Solinger (2017).

24. Roberts (1998).

25. Ward and Warren (2006).

26. In Rosalind Petchesky's *Abortion and Women's Choice*, she argues that abortion continues to represent the "fulcrum of a much broader ideological struggle in which the very meaning of the family, the state, motherhood, and young women's sexuality are contested" (1985, xi). See also Leslie Reagan's *Dangerous Pregnancies* (2010).

27. Collinicos (2015).

28. Propper, Burgess, and Gossage (2008); Propper, Burgess, and Green (2004).

29. Ross and Solinger specifically connect the rise of neoliberalism and pro-life conservative movements to white supremacy in their work on reproductive justice: "Business elites devoted to capital accumulation, [. . .] made a peculiar and enduring political alliance with economically and vulnerable whites hostile to racial equity and devoted to religious tradition" (2017, 100). Further, they write that the ideology of neoliberalism promotes a notion of motherhood that is characterized by whiteness and class privilege.

30. IOM (1990, 30).

31. IOM (1990, 53).

32. The 1990 IOM committee decided to define the standard for an ideal or "favorable" birth weight as six to eight pounds (1990, 4).

33. It was also not until the late 1970s that categories like Hispanic were created on the US Census; see discussion on this in chapter 3.

34. IOM (1990, 10). To create a more manageable and mobile standard, the 1990 report narrowed the definition of gestational weight gain to three main points: "weight just before delivery minus weight just before conception; total weight gain minus the infant's birth weight; rate per week, weight gained over a specified period divided by the duration of that period in weeks." The 1990 approach to maternal nutrition places gestational weight gain as the main measurable outcome, but only in the United States (IOM 1990, 13). Gestational weight gain is not a key standard in the United Kingdom. One practical issue for the lack of standardization in the United Kingdom is the fact that it does not routinely

weigh pregnant women. To establish gestational weight gain as a reliable measure, US doctors and scientists regularly collected and recorded weight during pregnancy.

35. Thompson (2007); Strings (2019).

36. Ludwick Fleck's work on the development of syphilis in the 1930s is a canonical example of what it means to study the production of knowledge by tracing a key object. For Fleck the production of knowledge is a collective effort, a "social activity," and social values can entrench or fix certain ideas regarding disease etiology, which influences the construction of facts (2008, 27). Fleck found that certain concepts are "traditional" or "normative" and resist change in the face of new and contradictory knowledge. He notes that contradictory information can be unseen or ignored in an effort to maintain traditional views. Similarly, Kuhn's (1996) work on paradigmatic shifts, referenced in chapter 1, is also relevant to this question of how science changes (or does not change). Kuhn and Fleck both examined the ways in which science is shaped by social and political environments. However, not everyone is equally included in this "social activity" of scientific knowledge production. And not all social ideologies have equal weight in shaping facts. My intervention here is to emphasize that regardless of epigenetic and DOHaD interventions, epistemic environments characterized by processes of racialization, like late liberalism and capitalism, maintain traditional or normative ideas of maternal health and responsibility.

37. The notion of weight gain/loss is derived from an older framing of the body as machine (Martin 1987).

38. As Mol and Law (2002) argue, the calorie is a unit of analysis based on social relations and requires laboratory practices, nutritional science, thermodynamics, and infrastructure like nutritional labeling (Dumit and de Laet 2014).

39. Landecker (2011).

40. Timmermans and Berg (2003); Latour (1987).

41. IOM (1990, 4).

42. See IOM (1990, 32) for the diagram of all potential variables of causation.

43. Studies published after the 1990 report emphasize that women who have high GWG can still deliver smaller than average babies, a phenomenon found across all socioeconomic levels in African American women (Collins et al. 2004). Recent research also suggests that women who are underweight and have low gestational weight gain can also have larger babies (Boney et al. 2005).

44. See IOM (1990, 33); Hill (1965).

45. Brownson et al. (2006).

46. The diagram on all potential determinants does include an environmental factor, but only defines it as "geography and climate" (IOM 1990, 32).

47. IOM (1990, 31).

48. IOM (1990, 32).

49. Saguey (2013, 46).

50. WHO (2000).

51. NICE (2010, 61; emphasis added).

52. NICE (2010, 20).

53. The UK report is titled *Weight Management before, during and after Pregnancy* (NICE 2010) The title of the report is confusing because although the report tells women to manage their weight before they get pregnant, it does not give weight gain recommendations during pregnancy. The report emphasizes that women should not lose more than half a kilogram of weight each week during pregnancy. The minimum weight limit is difficult to measure since the report also states that there is not enough evidence-based science to support the routine weighing of women during pregnancy. Therefore, women are not routinely weighed during pregnancy unless they have a BMI of over 30 and have been recommended to see a dietician.

54. There are exceptions and inconsistencies across different hospitals in the United Kingdom. Some will regularly weigh women who have a BMI of 35 or more, or 40 or more.

55. Rassmussen et al. (2009, 134). In chapters 4 and 6 of the same IOM report, scholars in the fields of epigenetics and DOHaD are cited, such as Gluckman and Hanson (2008), Barker (1995), and Waterland et al. (2008).

56. IOM and NRC (2009, 197).

57. The WHO (2002) defines *genetics* as the study of heredity and *genomics* as the study of genes and their mechanistic functions. The main difference is that genomics explores the relationship or interaction between and among genes to better understand the combined effect on development.

58. These observations are based off an online search of the committee member profiles.

59. IOM and NRC (2009, ix).

60. IOM and NRC (2009, 6).

61. IOM and NRC (2009, 71).

62. GWG is complexly associated with GDM, infant obesity, and preterm birth (Rasmussen, Catalano, and Yaktine 2009).

63. Siega-Riz et al. (2020).

## CHAPTER 3. POLITICS OF RECRUITMENT

1. I do not assume that the processes of implementing pregnancy trials are stable or consistent in different national contexts, which is why an emergent multisited ethnographic design enables and necessitates a capacious analysis (MacCormack and Strathern 1980).

2. See Wiles (1994). In casual conversations with a colleague, I asked, "What kinds of things did people say to you when you were pregnant?" The comments

varied, but most centered around the theme of fat shaming. An academic colleague told me that when she was pregnant and teaching at an elite university in the United States, she experienced horrible nausea. The only food that did not make her sick was orange popsicles. She was eating an orange popsicle on campus one day and her colleague commented, "Should you really be eating that right now?" My colleague felt totally embarrassed and ashamed.

3. This statistic came from a report written by a social scientist in charge of interviewing people who were approached to enroll in the UK trial. The report was never published.

4. This statistic is based on the unpublished UK trial report

5. Boero (2013b).

6. Berlant (1997); Bordo (2004).

7. Greenhalgh (2012); Kulick and Meneley (2005).

8. Saguy and Almeling (2008); Saguy and Riley (2005).

9. Puhl and Brownell (2006).

10. Often "good" health is causally associated with under or normal BMI, and overweight and obese classifications are automatically associated with "bad" health. This assumption creates blind spots in medical diagnoses. Someone who maintains their weight by eating a thousand calories of only cereal appears as normal weight and would not likely be asked to take a cholesterol panel at a regular checkup (Puhl and Heuer 2010).

11. Phelan et al. (2015).

12. Fat as a substance is not only socially complicated; it is also materially heterogenous. An article published in a scientific journal, "Can Brown Fat Win the Battle against White Fat?," explains that there are different kinds of fat. "Brown" fat is defined as "good" because it expels energy and releases heat, whereas "white" fat gathers around the waist and thighs, which is associated with adverse health outcomes. I spotlight this to emphasize how value-laden language is used to describe fat as a substance, which is indicative of how social and political meaning is enveloped in scientific explanations. To emphasize the urgency of this topic, the first sentence of the same article reads, "Obesity is rapidly emerging as the greatest challenge to human health worldwide." The extreme language employed here fuels a persistent focus on obesity (Elattar and Satyanarayana 2015, 2311).

13. Campos et al. (2006).

14. Gálvez (2018); Carney (2015); Hardin (2018); Dickinson (2019).

15. For instance, white women's bodies were framed as moral and healthy, while in contrast, Black women's bodies represented unhealthy excess (Strings 2019).

16. (Strings (2019, 6).

17. Strings (2019, 190).

18. Mansfield and Guthman (2015); Saldaña-Tejeda (2017).

19. Davin (1978).

20. Bridges (2011).

21. McPhail et al. (2016, 98).

22. Around one-third of all pregnancies (clinically recognized and unrecognized) end in miscarriages (Mukherjee et al. 2013).

23. I had access to collect data on the recruitment phase as a staff member in the trial. Once participants were enrolled in the trial, I would inform them that I was also collecting observations from their participation in the trial for my own ethnographic project.

24. I worked only at the West Coast site, but I participated in weekly meetings with the East Coast site.

25. Anderson, Borfitz, and Getz (2019); Fisher (2020).

26. The postpartum follow-ups included a separate amount that did not exceed $25.

27. Fisher (2020).

28. SmartStart protocol 2012

29. See NIH (1993) for policy and guidelines.

30. Another layer that challenged recruitment in diverse populations was the stigma that trials are not safe for pregnant women, and that losing weight or managing your diet is not "healthy" during pregnancy.

31. Pendergrass and Raji (2017).

32. SmartStart trial protocol, updated August 2012

33. See Nicol Valdez (2019).

34. Hunt and Megyesi (2008).

35. Mora (2014).

36. Schiebinger (2004); Mukhopadhyay (2008); Moses (2017); Saldaña-Tejeda (2017).

37. Epstein (2009). For example, see critiques of BiDil, a heart medication marketed to African Americans (Kahn 2012; Pollock 2012).

38. Fisher and Kalbaugh (2011, 2217).

39. Sullivan (2013).

40. Committee on Community-Based Solutions to Promote Health Equity in the United States et al. (2017).

41. I could not find the exact article the PI was referring to, and there is limited scholarship that documents the demographics of the principal investigators who design trials. However, based on historical underrepresentation and challenges in diversity and inclusion at elite levels of science, the majority of PIs in the United States are white men; some are women.

42. Clark et al. (2019).

43. Clarke et al. (2003); Rose (2006).

44. Gálvez (2018); Carney (2015).

45. Fessler and Rose (2019).

46 Nicol Valdez (2019).

47. Mexican American immigrant populations also pay taxes and volunteer as organ donors at higher rates than US-born white Americans (Chavez 2013)

48. Fisher (2020).

49. For more examinations on risk in trials see Cooper (2011); for risk and compensation in Phase I trials see Abadie (2010); Fisher (2020); for markets and experimentation see Petryna (2009).

50. Fisher (2009).

51. Fisher (2020). From a different perspective, but also reflective of neoliberal and capitalist environments that shape medical research and health care, is Abadie's (2010) ethnography *The Professional Guinea Pig*.

52. Petryna (2009).

53. United Kingdom Parliament (2021).

54. Propper, Burgess, and Gossage (2008).

55. Cooper and Waldby (2014).

56. McLeish and Redshaw (2019).

57. Material in this section is taken from Natali Valdez (2019a).

58. The project coordinator commented to me that the study's population included a greater percentage of African or Afro-Caribbean women relative to their percentage in the census population.

59. Epstein (2009).

60. Retnakaran et al. (2006).

61. This previously appeared in a commentary in Natali Valdez (2020).

62. Tompkins (2012); see Diana Burnett's piece in Gálvez, Carney, and Yates-Doerr (2020).


## CHAPTER 4. PREGNANT NARRATIVES

1. NICE (2010).

2. Data collection visit, glucose tolerance test (GTT), with H, May 12, 2014.

3. Ulijaszek and Lofink (2006).

4. Gálvez (2018, 6).

5. Paradies et al. (2015). The PI of the StandUp trial in the United Kingdom explained to me in an interview that only 30 percent of pregnant people with a BMI of 30 experience pregnancy complications, some of which are unrelated to their weight.

6. Carney (2015); Dickinson (2019); Yates-Doerr (2015); Hardin (2018).

7. Although bodies are technically larger since postwar industrialization, obesity rates are not currently increasing exponentially (CDC 2020).

8. See Mehta et al. (2011) for a public health perspective on the complexities across body image and pregnancy weight gain.

9. Black feminist scholars show that anti-Blackness permeates beauty and health standards, which change over time yet consistently perpetuate the notion that whiteness is the aesthetic ideal (Edu 2019).

10. Boero (2013a); Bordo (2004).

11. Greenhalgh and Carney (2014).

12. Fichter and Quadflieg (2016).

13. Wolf (2002); Martin (2008).

14. Moran-Thomas (2019).

15. Dodd et al. (2018).

16. Interview with HC, April 9, 2014.

17. In the same interview with the dietician, she emphasized that the intervention was designed to be simple and achievable for all women, regardless of ethnicity. She also noted that because they wanted to include a wide diversity of women in the trial, they did not add food diaries to the intervention, in case English was not someone's native tongue.

18. StandUp Trial (2012).

19. "A food with a high GI raises blood glucose more than a food with a medium or low GI" (American Diabetes Association 2015). The national diabetes organization in the United Kingdom states that "the glycemic index is a good way of making food choices, [and] glycemic load helps to work out how different sized portions of different foods compare with each other in terms of their blood glucose raising effect" (Diabetes.co.uk2019).

20. It is possible that in employing glycemic control during pregnancy, women may not gain as much weight because they are eating less sugar and saturated fat. Studies will often use low calorie interventions to improve glycemic control among adults with diabetes.

21. The postdoctoral fellows specialized in nutrition and behavioral interventions.

22. The aims of the intervention are clearly defined at the beginning of the handbook: "The programme combines advice on nutrition and physical activity with 'behavior change techniques' to help develop habits that encourage a healthier lifestyle with the ultimate aim of improving pregnancy outcome. The aims of the programme are for participants to: [. . .] Improve their blood sugar control during pregnancy; Learn how food and drink affects their blood sugar levels; Learn how to be physically active during pregnancy" (StandUp handbook).

23. StandUp Trial (2012).

24. StandUp Trial (2012) (emphasis added).

25. Thirty-five-week, data collection visit with H, June 4, 2014.

26. I am not sure what Slimming World recommends pregnant women should gain each week because there are no accepted guidelines in the United Kingdom. The Weight Watchers website gives a summary of the IOM guidelines that make recommendations for gestational weight gain.

27. Jean Nidetch from Queens, New York, started the Weight Watchers program in the early 1960s. She organized a group of women to meet at her home to "talk about how best to lose weight [. . .] . Today, that group of friends has grown to millions of women and men around the word" ("Weight Watchers" n.d.).

28. Intervention session with D, May 3, 2014.

29. The division of labor in the StandUp trial was organized by the NHS through levels or "bands." The bands refer to the amount of skill and pay required for each job within the NHS. For instance, the principal investigator was band 9, which includes all senior professionals with a doctoral degree and advanced research experience. Research midwives were band 6 or 5 depending on experience. Research assistants were usually band 4 and had the potential to get promoted, depending on work experience and earned degrees or certificates. The health trainers were band 3 and could not be promoted in band level or pay. Band 3 did not include any midwives and they were paid the least compared to the other staff; they were also solely in charge of delivering the intervention. In interviews with health trainers, they commented to me that they often felt unappreciated and wanted more guidance from superior staff. The bands, salary, and promotion options reflect hierarchical labor structure.

30. Sheryl lives and works in Scotland, which is relevant because of a cultural opinion that frames "the north" as less sophisticated than "the south."

31. Bretan (2017).

32. Interview with SH, April 22, 2014.

33. One of the surgical interventions I learned of during trial implementation was a sleeve gastrectomy. A "sleeve" is a surgery that removes more than half the stomach to reduce the amount of food a person can eat and digest. If you eat too much with a "sleeve," you experience intense pain.

34. Intervention session, with S, May 3, 2014.

35. Burnett (2014).

36. Keys (1980)

37. Tompkins (2012, 6).

38. Duster (2005); TallBear (2013); Roberts (2012).

39. Intervention session with DB, May 16, 2014.

40. Intervention session with A, May 9, 2014.

41. Greenhalgh (2012); Greenhalgh and Carney (2014).

42. Mol (2013).

43. Suzuki et al. (2016).

44. Richardson (2015, 224–25).

45. Bollati and Baccarelli (2010); Weaver, Meaney, and Szyf (2006).

46. Richardson (2015, 224).

47. (Richardson et al. 2014; Sharp, Lawlor, and Richardson 2018).

48. Sharp, Lawlor, and Richardson (2018).

49. Scott et al. (2014).

50. Athearn et al. (2004).

51. Throughout this chapter I use the term *women*, because all of the participants I encountered identified as cis-women.

52. Colen (1995).

53. This message has persisted in the Euro-American context since the inception of population health (Foucault 1994a; Davin 1978).

54. Ginsburg and Rapp (1991); Clarke and Haraway (2018).


CHAPTER 5. ENVIRONMENTAL ANIMATIONS

1. Barker et al. (1993).

2. Meloni and Testa (2014).

3. Jablonka and Lamb (1995, 2006).

4. Pickersgill et al. (2013).

5. See Shostak and Moinester (2015); Lappé and Landecker (2015); Lamoreaux (2016); See also Niewohner's work on "spatio-temporal contexts" (2011, 290).

6. Landecker and Panofsky (2013).

7. Landecker (2010) explains that in the twentieth century the emergence
of nutrigenetics focused on the ways in which bodies absorb nutrition and the
molecular interactions involved in metabolism. In contrast, nutrition in an epigenetic age is framed as a problem of gene expression in critical periods with
potential effect on future generations.

8. Although this aspect of intergenerational inheritance is hotly debated, the
narratives around food as environmental exposures are mobilized as if it were
a concrete reality. DOHaD theories claim that nutritional exposures can impact
health across lifespans.

9. Weaver et al. (2006); Szyf (2009); Yehuda et al. (2016).

10. Darling et al. (2016).

11. NIEHS (2017).

12. Murphy reminds us that "all people alive today contain PCBs (polychlori-
nated biphenyls)," which are pollutants that circulate globally as a result of the
massive industrialization of our planet in the last two hundred years (2017b, 494).

13. Lappé, Hein, and Landecker (2019).

14. Maher, Fraser, and Wright (2010); Warin et al. (2012).

15. Kenney and Müller (2017); Singh (2012).

16. Manderson (2016, 154).

17. El-Haj (2007); Mansfield and Guthman (2015).

18. In order to maintain anonymity I cannot cite the PI's name.

19. Kenney and Müller (2017).

20. Weaver, Meaney, and Szyf (2006).

21. Haraway (1997); Freccero (2011); Thompson (2013).

22. Muller and Kenny (2016, 31).

23. Weaver, Meaney, and Szyf (2006).

24. Duden (1993); Casper (1998).

25. Mullings and Wali (2001, 2).

26. Agard-Jones (2013); Alaimo (2016); Murphy (2017b); Sasser (2018).

27. See Natali Valdez (2019).

28. See Haraway (1988, 1990a, 1990b, 1997); Mamo and Fosket (2009); and more recently Yates-Doerr (2020).

29. Non (2014); Thayer and Non (2015).

30. Non et al. (2021).

31. Sugden et al. (2020).

32. In June 2020, the death toll from the COVID-19 pandemic was fifty times the number of people who died on 9/11, or equivalent to a jumbo airplane crashing every single day for three months (Dale and Stylianou 2020).

## CHAPTER 6. PROSPECTING PREGNANCIES

1. Data collection visit, GTT, May 4, 2014.

2. This increased production is framed as a "surplus" for extraction in the trial (Hytten 1985).

3. Even though the participant consented to this test at the enrollment, it is common for women to forget the purpose of the glucose tolerance test.

4. I observed more than twenty data collection visits in which the UK staff members had to collect blood. I paid close attention to how blood is collected because the published papers of RCTs do not discuss this time and labor-intensive process.

5. See Hayden (2003) for an understanding of bioprospecting in Mexico. See also Alberto Morales's (2019) dissertation for bioprospecting in relation to economies of knowledge. My approach draws on Marxist, postcolonial, and ecofeminist perspectives that emphasize how environments and women's labor are exploited by similar power structures like capitalism and patriarchy. Unlike traditional frameworks of ecofeminism of the 1970s, I draw on a queer feminist materialist stance on the relationship between women and nature: women's bodies are not inherently closer to nature because they are deemed "female" or can reproduce. My approach focuses on how epistemes of racial capitalism and patriarchy create the conditions of exploitation that target certain gendered and raced bodies and environments.

6. The term *pregnant biobits* draws on existing feminist work. See, for instance, Wendy Chavkin's work for the term *body bits* (Chavkin and Maher 2010); see also Farquhar and Lock (2007); Rapp (2011).

7. Lock reminds us that the circulation of bodies has a long history across science, religion, and wars. Body parts have been valuable to shame enemies, for

religious rituals, and for medical and capitalist purposes (Farquhar and Lock 2007, 571). See also Haraway (1997, 142).

8. From a consortium meeting on the "tally for SHARED Measures" in 2013.

9. However, she mentioned that she would not be collecting breastmilk specifically. The breastmilk reflected a data and material boundary for the PI.

10. Fannin and Kent (2015).

11. Even though it was imagined as easier to standardize biosamples rather than behavioral data, both the US and the UK trials struggled with the standardization, collection, and storage of biosamples. After one of the many meetings about the standardization of biosample collection, the US PI stated to me, "Everyone thinks that collecting biosamples is easier, more standardized, but I think it is just as hard as collecting behavior data." The US PI understood that the standardization of biological samples also required human behavioral coordination.

12. Darling et al. (2016).

13. Kelishadi, Badiee, and Adeli (2007, 518).

14. Lappé and Landecker (2015, 9).

15. For instance, the processes of extracting pregnant data are similar to strategies in surveillance capitalism, in that these data are treated as surplus or "free" in exchange for perceived benefits of participating in the trials. For some participants, the trial provided extra health-care services. However, one main distinction is scale: unlike the collection of millions of "clicks" and "likes" per second that is characteristic of what Zuboff (2019) calls surveillance capitalism, the collection of biobehavioral data in trials requires more time and labor. Another key divergence is that all behavioral experiences emerge as valuable in surveillance capitalism, whereas only certain kinds of behavioral experiences count as valuable data for the trials. One reason for this is that trials are limited in their capacity to collect and analyze data.

16. Halberstam (2005).

17. See Marx (1976, ch. 10).

18. There are different kinds of temporalities beyond the Western capitalist imaginary of time (Blaser 2013).

19. Szyf (2009).

20. Landecker and Panofsky (2013, 339; emphasis added).The unpredictability and nondeterminate temporality inherent to stochasticism are similar to the ways in which Nietzsche, Bergson, Deleuze, and Irigaray frame temporality as open, becoming, uncertain, unpredictable, and nonlinear (Grosz 2004).

21. Yehuda et al. (2016).

22. Epigenetic time, in particular scientific imaginaries, has a transversal character, meaning that it lies across the past, present, and future. To be sure, this transversality has some direction. Based on current metaphysics, the future may not impact the past as much as the past influences the present and future.

23. SmartStart (2012, 9).

24. Exact quote from the protocol: "Based on first trimester measured weight and measured height. [. . .] If the earliest weight measurement is conducted at 14 weeks 0 days to [. . .] 15 weeks 6 days, 1lb or 2lb will be subtracted from the measured weight respectively, to adjust to a first trimester weight" (SmartStart 2012, 9).

25. Butt et al. (2014).

26. Butt et al. (2014, 173).

27. Interview with SmartStart staff member, July 12, 2013.

28. Informal conversation with ABB after coordinator meeting with the research coordinating unit, July 11, 2013.

29. StandUp Trial (2012, 22).

30. Rapp (2011); Casper (1998).

31. Kaempf et al. (2006).

32. WHO (2001).

33. See the discussion of the Biomarkers Definition Working Group in Singh and Rose (2009).

34. For instance, "the use of biomarkers in basic and clinical research as well as in clinical practice has become so commonplace that their presence as primary endpoints in clinical trials is now accepted almost without question" (Strimbu and Tavel 2010, 463).

35. El-Haj (2007).

36. Singh and Rose (2009).

37. See Lock's (2013) concept of somatic reductionism.

38. Even though the trial intended to recruit the father's or partners into the study for some data collection, it was not successful in collecting enough samples.

39. I met Connie first in 2012 and again in 2014 during my longer data fieldwork phase.

40. This does not include the cord blood and placental tissue samples, which required a separate set of processing at delivery.

41. Cytokines are an ambiguous substance, and they are difficult to classify because they have multiple molecular roles (Dinarello 2007).

42. Agard-Jones (2013).

43. KHP Conference, June 18, 2014 (my emphasis).

44. Buchanan and Xiang (2005).

45. For instance, it was important to collect the maternal blood samples during the glucose tolerance test at twenty-eight weeks (as the PI mentioned, the cytokines needed to be evaluated at twenty-eight weeks) because those samples could be used to find a biomarker associated with GDM.

46. It was extremely challenging to standardize the oral glucose tolerance tests across the US and UK health-care landscapes. Both the United States and United Kingdom used the more conservative two-hour 75g OGTT based on the HAPO study (Coustan et al. 2010). Since it is a more conservative test, more people are diagnosed as a result. However, the glucose tolerance test varied widely within

standard prenatal care in both the United States and United Kingdom. Committee on Practice Bulletins—Obstetrics (2018). One in ten people in the United States and United Kingdom have diabetes. Mexican and Black Americans are at higher risk of diabetes (Montoya 2011), and Asian and South Asian populations in the United Kingdom are at high risk (Diabetes UK n.d.).

47. Farrar et al. (2016).

48. Chen et al. (2015).

49. Pantell et al. (2019).

50. CDC (2019). A main challenge in the health-care landscape is the lack of standardized diagnosing. There is no medical consensus on whether to use a one- or two-hour test, or whether all pregnant people should take the test as part of standard prenatal care. Even associations with BMI categories are inconsistent. In some places having a BMI of 25, 30, or more would entail different kinds of OGTTs or random blood sugar tests. In the United Kingdom, pregnant participants usually receive a random blood sugar test; only if these results are too high are they recommended for an OGTT, and usually most NHS hospitals only complete a one-hour test.

Both trials followed the most conservative guidelines for GDM diagnosis, which draws on the Hyperglycemia and Adverse Pregnancy Outcome (HAPO) study. Based on these guidelines, a negative test, meaning the absence of gestational diabetes mellitus, requires that the baseline blood sugar levels be at or less than 5.1; at the one-hour mark sugar levels should be at or above 10, and at the two-hour mark the sugar levels should be less than 7.8.

51. Still, drawing on capitalist logics, investing in affordable housing can also be profitable in the present and future. For instance, redistributing the funds of one lifestyle pregnancy trial (about $2–5 million depending on sample size) to houseless pregnant populations can make an impact on health outcomes in the present and future.

52. Martin (2007).

## CONCLUSION

1. For another aspect of foreclosure that is connected to the larger movement of the molecularization of life, see reference to Darling et al. (2016) in chapter 1; and for the notion of somatic reductionism see Lock (2013).

2. This is a global issue, but most of the top journals are published in the United States or Europe.

3. Nelson et al. (2021).

4. "Reality Check on Reproducibility" (2016).

5. There is limited data on the reproducibility of lifestyle interventions; however, the existing literature on the reproducibility of clinical research writ large points to the same issue (Niven et al. 2018).

6. See James Ferguson's (1994) work on "failed" NGO programs.

7. The PI explained further that "some studies are showing that actually weight loss during gestation may be necessary."

8. Siega-Riz et al. (2020).

9. There is one main study that has followed up with children at three years of age, a similarly designed intervention; the results of the follow-up study were also inconclusive (Dodd et al. 2018).

10. Nuru-Jeter et al. (2018).

11. Salway et al. (2016).

12. Alexander (2012); Roberts (1998, 2009, 2012).

13. Davis (2019).

14. I spoke with some of the staff, and they said this study is challenging because it requires them to go into participants' homes; similar to the SmartStart trial, the main focus is on the weight of the toddlers.

15. Meloni (2017).

16. Davis (2019); Edu (2019).

17. Yehuda (2015).

18. See Pico (2016); *Dear White People* (2018).

19. Pentecost and Meloni (2020).

20. Jasanoff (2007); Subramaniam and Herzig (2019); Reardon (2004, 2017); Roy (2018); Fullwiley (2011).

21. What I suggest is a more specific attention to the social, which is not to be mistaken for an uncritical return to social constructionism. Thinking with the material and social together is vital, yet our social and material accounts can be more precise in disentangling and making accountable racist, sexist, and capitalist forces.

22. Waggoner (2017).

23. Niewöhner (2011).

24. My approach draws on a long legacy of feminist and postmodern, post-structuralist scholars who analyze the gendered or racialized structures of language; see the canonical essay by Emily Martin (1991), "The Sperm and the Egg."

25. See the work of Indigenous scholars like Million (2013); Sasser (2014); and Lappé, Hein, and Landecker (2019). See also Michele Murphy's working groups and collaborations at https://michellemurphy.net/collaborations/.

## EPILOGUE

The title is a riff or sampling of Emily Dickinson's poem title "Forever Is Composed of Nows."

1. These were just a few of the names that emerged in the news coverage; there were many other murders in the month of May, including members of the transgender community like Nina Pop, a Black woman killed in Missouri.

2. CDC (2021a); *New York Times* (2021).

3. Some estimations claim that African Americans and Hispanics make up closer to 30 percent of all clinical trials funded by the NIH (Fisher and Kalbaugh 2011, 2217).

4. Fisher (2020, 4).

5. Fisher (2020, 5).

6. "AAPA Statement on Race & Racism" (n.d.).

7. For instance, a recent NIH project titled "Epigenome and Gene Expression Signatures of Racial Differences in Chronic Low Back Pain" aims "to uncover novel epigenetic and gene expression mechanisms that underlie racial differences." This focus still adheres to understanding racial difference through epigenetics, rather than understanding differences in health outcomes in relation to racism (NIH 2021).

8. Berlant (2011).

# References

"AAPA Statement on Race & Racism." n.d. Accessed July 3, 2019. http://physanth
.org/about/position-statements/aapa-statement-race-and-racism-2019/.

Abadie, Roberto. 2010. *The Professional Guinea Pig: Big Pharma and the Risky
World of Human Subjects.* Durham, NC: Duke University Press.

Abu-Lughod, Lila. 1990. "Can There Be a Feminist Ethnography?" *Women &
Performance: A Journal of Feminist Theory* 5 (1): 7–27. https://doi.org/10
.1080/07407709008571138.

———. 2000. "Locating Ethnography." *Ethnography* 1 (2): 261–67. https://doi.org
/10.1177/14661380022230778.

Ackerman, Sara L., Katherine Weatherford Darling, Sandra Soo-Jin Lee, Robert
A. Hiatt, and Janet K. Shim. 2016. "Accounting for Complexity: Gene–
Environment Interaction Research and the Moral Economy of Quantifica-
tion." *Science, Technology, & Human Values* 41 (2): 194–218. https://doi.org
/10.1177/0162243915595462.

Adams, Vincanne, Michelle Murphy, and Adele E Clarke. 2009. "Anticipation:
Technoscience, Life, Affect, Temporality." *Subjectivity* 28 (1): 246–65. https://
doi.org/10.1057/sub.2009.18.

Agard-Jones, Vanessa. 2013. "Bodies in the System." *Small Axe: A Caribbean
Journal of Criticism* 17 (3): 182–92. https://doi.org/10.1215/07990537
-2378991.

Alaimo, Stacy. 2016. *Exposed: Environmental Politics and Pleasures in Post-
human Times.* 1st ed. Minneapolis: University of Minnesota Press.

Alcoff, Linda. 2015. *The Future of Whiteness*. Cambridge, UK: Polity Press.

Alexander, Michelle. 2012. *The New Jim Crow: Mass Incarceration in the Age of Colorblindness*. New York: The New Press.

Allen, Theodore W., and Jeffrey B. Perry. 2012. *The Invention of the White Race*. Vol. 1, *Racial Oppression and Social Control*. 2nd ed. London: Verso.

Almeling, Rene. 2020. *GUYnecology: The Missing Science of Men's Reproductive Health*. 1st ed. Oakland: University of California Press.

Almond, Douglas, and Janet Currie. 2011. "Killing Me Softly: The Fetal Origins Hypothesis." *Journal of Economic Perspectives: A Journal of the American Economic Association* 25 (3): 153–72. https://doi.org/10.1257/jep.25.3.153.

American Diabetes Association. 2015. "Understanding Carbohydrates." http://archives.diabetes.org/food-and-fitness/food/what-can-i-eat/understanding-carbohydrates/.

Anderson, Warwick. 2006. *Colonial Pathologies: American Tropical Medicine, Race, and Hygiene in the Philippines*. Durham, NC: Duke University Press.

Anderson, Annick, Deborah Borfitz, and Kenneth Getz. 2019. "Differences in Clinical Research Perceptions and Experiences by Age Subgroup." *Therapeutic Innovation & Regulatory Science* (January). https://doi.org/10.1177/2168479018814723.

Anzaldúa, Gloria. 2012. *Borderlands: La Frontera: The New Mestiza*. 4th ed. San Francisco: Aunt Lute Books.

Appel, Hannah. 2019. *The Licit Life of Capitalism: U.S. Oil in Equatorial Guinea*. Durham, NC: Duke University Press.

Arendt, Hannah. 1973. *The Origins of Totalitarianism*. 1st ed. New York: Harcourt, Brace, Jovanovich.

Artiga, Samantha, Jennifer Kates, Josh Michaud, and Latoya Hill. 2021. "Racial Diversity within COVID-19 Vaccine Clinical Trials: Key Questions and Answers." KFF. January 26. www.kff.org/racial-equity-and-health-policy/issue-brief/racial-diversity-within-covid-19-vaccine-clinical-trials-key-questions-and-answers/.

Atanasoski, Neda, and Kalindi Vora. 2019. *Surrogate Humanity: Race, Robots, and the Politics of Technological Futures*. Durham, NC: Duke University Press.

Athearn, Prudence N., Patricia A. Kendall, Virginia Hillers, Mary Schroeder, Verna Bergmann, Gang Chen, and Lydia C. Medeiros. 2004. "Awareness and Acceptance of Current Food Safety Recommendations During Pregnancy." *Maternal and Child Health Journal* 8 (3): 149–62. https://doi.org/10.1023/B:MACI.0000037648.86387.1d.

Bailey, Moya. 2021. *Misogynoir Transformed: Black Women's Digital Resistance*. New York: New York University Press.

Bailey, Moya, and Whitney Peoples. 2017. "Towards a Black Feminist Health Science Studies." *Catalyst: Feminism, Theory, Technoscience* 3 (2). https://go

.gale.com/ps/i.do?p=AONE&sw=w&issn=23803312&v=2.1&it=r&id=GALE %7CA561685864&sid=googleScholar&linkaccess=abs.

Banton, M. 1977. *The Idea of Race*. London: Tavistock.

Baptist, Edward E. 2016. *The Half Has Never Been Told: Slavery and the Making of American Capitalism*. Reprint ed. New York: Basic Books.

Barker, D. J., and C. Osmond. 1986. "Infant Mortality, Childhood Nutrition, and Ischaemic Heart Disease in England and Wales." *Lancet* 1 (8489): 1077–81. https://doi.org/10.1016/s0140-6736(86)91340-1.

Barker, D. J. P. 1995. "Fetal Origins of Coronary Heart Disease." *BMJ* 311 (6998): 171–74. https://doi.org/10.1136/bmj.311.6998.171.

———. 2007. "The Origins of the Developmental Origins Theory." *Journal of Internal Medicine* 261 (5): 412–17. https://doi.org/10.1111/j.1365-2796.2007.01809.x.

Barker, D. J. P., K. M. Godfrey, P. D. Gluckman, J. E. Harding, J. A. Owens, and J. S. Robinson. 1993. "Fetal Nutrition and Cardiovascular Disease in Adult Life." *Lancet* 341 (8850): 938–41. https://doi.org/10.1016/0140-6736(93)91224-A.

Barker, D. J. P., J. E. Harding, J. A. Owens, J. S. Robinson, et al. 1993. "Fetal Nutrition and Cardiovascular Disease in Adult Life." *Lancet* 341 (8850): 938–41.

Barnes, Barry, and John Dupré. 2008. *Genomes and What to Make of Them*. 1st ed. Chicago: University of Chicago Press.

Bell, J. 2010. "A Historical Overview of Preeclampsia-Eclampsia." *Journal of Obstetric, Gynecologic, & Neonatal Nursing* 39 (5): 510–18.

Benjamin, Ruha. 2013. *People's Science: Bodies and Rights on the Stem Cell Frontier*. 1st ed. Stanford, CA: Stanford University Press.

———. 2019. *Race after Technology: Abolitionist Tools for the New Jim Code*. Medford, MA: Polity.

Berlant, Lauren. 1997. *The Queen of America Goes to Washington City: Essays on Sex and Citizenship*. Durham, NC: Duke University Press.

———. 2011. *Cruel Optimism*. Durham, NC: Duke University Press.

Bird, Adrian. 2002. "DNA Methylation Patterns and Epigenetic Memory." *Genes & Development* 16 (1): 6–21. https://doi.org/10.1101/gad.947102.

Blaser, M. 2013 "Ontological Conflicts and the Stories of Peoples in Spite of Europe: Towards a Conversation on Political Ontology." *Current Anthropology* 54 (5): 547–68.

Boddy, A., A. Fortunato, M. Syres, and A. Aktipis. 2015. "Fetal Microchimerism and Maternal Health: A Review and Evolutionary Analysis of Cooperation and Conflict Beyond the Womb." *BioEssays* 37 (10): 1106–18.

Boddy, J. 1998. "Remembering Amal: On Birth and the British in Northern Sudan." In *Pragmatic Women and Body Politics*, edited by M. Lock and P. A. Kaufert, 28–57. New York: Cambridge University Press.

Boero, Natalie. 2013a. *Killer Fat: Media, Medicine, and Morals in the American "Obesity Epidemic"*. New Brunswick, NJ: Rutgers University Press.

——. 2013b. "Obesity in the Media: Social Science Weighs In." *Critical Public Health* 23 (3): 371–80. https://doi.org/10.1080/09581596.2013.783686.

Bollati, Valentina, and Andrea Baccarelli. 2010. "Environmental Epigenetics." *Heredity* 105 (1): 105–12. https://doi.org/10.1038/hdy.2010.2.

Bonasio, R., Shengjiang Tu, and Danny Reinberg. 2010. "Molecular Signals of Epigenetic States." *Science* 330: 612–16.

Boney, C. M., A. Verma, R. Tucker, and B. R. Vohr. 2005. "Metabolic Syndrome in Childhood: Association with Birth Weight, Maternal Obesity, and Gestational Diabetes Mellitus." *Pediatrics* 115 (3): e290–96.

Bordo, Susan. 2004. *Unbearable Weight: Feminism, Western Culture, and the Body*. 2nd ed., 10th anniv. ed. Berkeley: University of California Press.

Bowker, G., and S. Star. 1999. *Sorting Things Out: Classification and Its Consequences*. Cambridge, MA: MIT Press.

Bretan, Juliette. 2017. "The North-South Divide Is Getting Worse—Because London Is Sucking All Opportunity from the Rest of the UK." *Independent*, August 11. www.independent.co.uk/voices/north-south-divide-life-expectancy-northern-powerhouse-inequality-opportunities-a7887771.html.

Brewer, T., and S. Brewer. 1983. *The Brewer Medical Diet for Normal and High-Risk Pregnancy*. New York: Simon and Schuster.

Bridges, Khiara M. 2011. *Reproducing Race: An Ethnography of Pregnancy as a Site of Racialization*. 1st ed. Berkeley: University of California Press.

Briggs, Helen. 2014. "Pre-Pregnancy Diet 'Affects Genes.'" *BBC News*, April 30, sec. Health. www.bbc.com/news/health-27211153.

Briggs, Laura. 2017. *How All Politics Became Reproductive Politics: From Welfare Reform to Foreclosure to Trump*. 1st ed. Oakland: University of California Press.

Brown, Wendy. 2019. *In the Ruins of Neoliberalism: The Rise of Antidemocratic Politics in the West*. New York: Columbia University Press.

Browne, Simone. 2015. *Dark Matters: On the Surveillance of Blackness*. Durham, NC: Duke University Press.

Brownson, Ross C., Matthew W. Kreuter, Barbara A. Arrington, and William R. True. 2006. "Translating Scientific Discoveries into Public Health Action: How Can Schools of Public Health Move Us Forward?" *Public Health Reports* 121 (1): 97–103.

Brugge, Doug, and Rob Goble. 2002. "The History of Uranium Mining and the Navajo People." *American Journal of Public Health* 92 (9): 1410–19.

Buchanan, Thomas A., and Anny H. Xiang. 2005. "Gestational Diabetes Mellitus." *Journal of Clinical Investigation* 115 (3): 485–91. https://doi.org/10.1172/JCI24531.

Burnett, Diana A. 2014. "Utilizing Photo-Elicitation to Explore the Impact of the Nutrition Transition on the Consumption Patterns, Lifestyle Practices, and Health of Black Women in Salvador da Bahia, Brazil." Unpublished manuscript, University of Pennsylvania.

Butt, Kimberly, Ken Lim, Stephen Bly, Kimberly Butt, Yvonne Cargill, Greg Davies, et al. 2014. "Determination of Gestational Age by Ultrasound." *Journal of Obstetrics and Gynaecology Canada* 36 (2): 171–81. https://doi.org/10.1016 /S1701-2163(15)30664-2.

Bygren, Lars Olov, Petter Tinghög, John Carstensen, Sören Edvinsson, Gunnar Kaati, Marcus E. Pembrey, and Michael Sjöström. 2014. "Change in Paternal Grandmothers' Early Food Supply Influenced Cardiovascular Mortality of the Female Grandchildren." *BMC Genetics* 15 (1): 12. https://doi.org/10.1186/1471 -2156-15-12.

Campos, Paul, Abigail Saguy, Paul Ernsberger, Eric Oliver, and Glenn Gaesser. 2006. "The Epidemiology of Overweight and Obesity: Public Health Crisis or Moral Panic?" *International Journal of Epidemiology* 35 (1): 55–60. https:// doi.org/10.1093/ije/dyi254.

Carney, Megan A. 2015. *The Unending Hunger: Tracing Women and Food Insecurity across Borders.* 1st ed. Oakland: University of California Press.

Cartwright, Nancy, and Jeremy Hardie. 2012. *Evidence-Based Policy: A Practical Guide to Doing It Better.* Oxford: Oxford University Press.

Casper, Monica J. 1998. *The Making of the Unborn Patient: A Social Anatomy of Fetal Surgery.* New Brunswick, NJ: Rutgers University Press.

Centers for Disease Control and Prevention (CDC). 2019. "Type 2 Diabetes." June 11. www.cdc.gov/diabetes/basics/type2.html.

———. 2020. "Obesity Is a Common, Serious, and Costly Disease." June 29. www .cdc.gov/obesity/data/adult.html.

———. 2021a. "COVID-19 Mortality Overview." https://www.cdc.gov/nchs/covid19 /mortality-overview.htm.

———. 2021b. "Health Equity Considerations and Racial and Ethnic Minority Groups." April 19. www.cdc.gov/coronavirus/2019-ncov/community/health -equity/race-ethnicity.html.

Central Intelligence Agency. 2015. *The World Factbook.* https://www.cia.gov/the -world-factbook.

Chavez, Leo. 2013. *The Latino Threat: Constructing Immigrants, Citizens, and the Nation, Second Edition.* 2nd ed. Stanford, CA: Stanford University Press.

Chavkin, Wendy, and JaneMaree Maher, eds. 2010. *The Globalization of Motherhood: Deconstructions and Reconstructions of Biology and Care.* New York: Routledge.

Chen, Mel Y. 2012. *Animacies: Biopolitics, Racial Mattering, and Queer Affect.* Durham, NC: Duke University Press.

Chen, Qi, J. Wei, M. Tong, L. Yu, A. C. Lee, Y. F. Gao, and M. Zhao. 2015. "Associations between Body Mass Index and Maternal Weight Gain on the Delivery of LGA Infants in Chinese Women with Gestational Diabetes Mellitus." *Journal of Diabetes and Its Complications* 29 (8): 1037–41. https://doi.org/10.1016/j.jdiacomp.2015.08.017.

Clark, Luther T., Laurence Watkins, Ileana L. Piña, Mary Elmer, Ola Akinboboye, Millicent Gorham, Brenda Jamerson, et al. 2019. "Increasing Diversity in Clinical Trials: Overcoming Critical Barriers." *Current Problems in Cardiology* 44 (5): 148–72. https://doi.org/10.1016/j.cpcardiol.2018.11.002.

Clarke, Adele, and Donna J. Haraway, eds. 2018. *Making Kin Not Population: Reconceiving Generations*. 1st ed. Chicago: Prickly Paradigm Press.

Clarke, Adele E., Laura Mamo, Jennifer R. Fishman, Janet K. Shim, and Jennifer Ruth Fosket. 2003. "Biomedicalization: Technoscientific Transformations of Health, Illness, and U.S. Biomedicine." *American Sociological Review* 68 (2): 161. https://doi.org/10.2307/1519765.

Clough, Patricia, and Craig Willse. 2011. *Beyond Biopolitics: Essays on the Governance of Life and Death*. Durham, NC: Duke University Press.

Cohen, Ed. 2009. *A Body Worth Defending: Immunity, Biopolitics, and the Apotheosis of the Modern Body*. Durham, NC: Duke University Press.

Cohn, Simon, Megan Clinch, Chris Bunn, and Paul Stronge. 2013. "Entangled Complexity: Why Complex Interventions Are Just Not Complicated Enough." *Journal of Health Services Research & Policy* 18 (1): 40–43. https://doi.org/10.1258/jhsrp.2012.012036.

Colen, S. 1995. "Like a Mother to Them: Stratified Reproduction and West Indian Childcare Workers and Employers in New York." In *Conceiving the New World Order: The Global Politics of Reproduction*, edited by Faye D. Ginsburg and Rayna Rapp, 78–102. Berkeley: University of California Press.

Collinicos, Alex. 2015. "Britain and the Crisis of the Neoliberal State." *International Socialism Journal* (145). http://isj.org.uk/britain-and-the-crisis-of-the-neoliberal-state/.

Collins, James W., Richard J. David, Arden Handler, Stephen Wall, and Steven Andes. 2004. "Very Low Birthweight in African American Infants: The Role of Maternal Exposure to Interpersonal Racial Discrimination." *American Journal of Public Health* 94 (12): 2132–38. https://doi.org/10.2105/AJPH.94.12.2132.

Committee on Community-Based Solutions to Promote Health Equity in the United States, Board on Population Health and Public Health Practice, Health and Medicine Division, and National Academies of Sciences, Engineering, and Medicine. 2017. *Communities in Action: Pathways to Health Equity*. Edited by James N. Weinstein, Amy Geller, Yamrot Negussie, and Alina Baciu. Washington, DC: National Academies Press. https://doi.org/10.17226/24624.

Committee on Practice Bulletins—Obstetrics. 2018. "ACOG Practice Bulletin No. 190: Gestational Diabetes Mellitus." *Obstetrics and Gynecology* 131 (2): e49–64. https://doi.org/10.1097/AOG.0000000000002501.

Cooper, Melinda. 2008. *Life as Surplus: Biotechnology and Capitalism in the Neoliberal Era*. Seattle: University of Washington Press.

———. 2011. "Trial by Accident: Tort Law, Industrial Risks and the History of Medical Experiment." *Journal of Cultural Economy* 4 (1): 81–96. https://doi.org/10.1080/17530350.2011.535374.

Cooper, Melinda, and Catherine Waldby. 2014. *Clinical Labor: Tissue Donors and Research Subjects in the Global Bioeconomy*. Durham, NC: Duke University Press.

Coustan, Donald R., Lynn P. Lowe, Boyd E. Metzger, and Alan R. Dyer. 2010. "The Hyperglycemia and Adverse Pregnancy Outcome (HAPO) Study: Paving the Way for New Diagnostic Criteria for Gestational Diabetes Mellitus." *American Journal of Obstetrics and Gynecology* 202 (6) (June 1, 2010): 654.e1–e6. https://doi.org/10.1016/j.ajog.2010.04.006.

Cromer, Risa. 2018. "Saving Embryos in Stem Cell Science and Christian Adoption." *New Genetics and Society* 37 (4):362–86. https://doi.org/10.1080/14636778.2018.1546574.

———. 2019. "Making the Ethnic Embryo: Enacting Race in US Embryo Adoption." *Medical Anthropology* 38 (7): 603–19. https://doi.org/10.1080/01459740.2019.1591394.

———. 2020. "'Our Family Picture Is a Little Hint of Heaven': Race, Religion, and Selective Reproduction in US 'Embryo Adoption.'" *Reproductive Biomedicine and Society Online* 11: 9–17. https://doi.org/10.1016/j.rbms.2020.08.002.

Dale, Becky, and Nassos Stylianou. 2020. "What Is the True Death Toll of the Coronavirus Pandemic?" *BBC News*, June 18, sec. World. www.bbc.com/news/world-53073046.

Darling, Katherine Weatherford, Sara L. Ackerman, Robert H. Hiatt, Sandra Soo-Jin Lee, and Janet K. Shim. 2016. "Enacting the Molecular Imperative: How Gene-Environment Interaction Research Links Bodies and Environments in the Post-Genomic Age." *Social Science & Medicine* 155 (April): 51–60. https://doi.org/10.1016/j.socscimed.2016.03.007.

Davin, Anna. 1978. "Imperialism and Motherhood." *History Workshop Journal* 5 (1): 9–66. https://doi.org/10.1093/hwj/5.1.9.

Davis, Dána-Ain. 2019. *Reproductive Injustice: Racism, Pregnancy, and Premature Birth*. New York: New York University Press.

*Dear White People*. 2018. Season 2, chapter 8.

Deomampo, Daisy. 2016. *Transnational Reproduction: Race, Kinship, and Commercial Surrogacy in India*. Reprint ed. New York: New York University Press.

———. 2019. "Racialized Commodities: Race and Value in Human Egg Donation." *Medical Anthropology* 38 (7): 620–33. https://doi.org/10.1080/01459740.2019.1570188.

Desmond, Matthew. 2019. "American Capitalism Is Brutal: You Can Trace That to the Plantation." *New York Times*, August 14, sec. Magazine. www.nytimes.com/interactive/2019/08/14/magazine/slavery-capitalism.html

Diabetes.co.uk. 2019. "Glycemic Load." www.diabetes.co.uk/diet/glycemic-load.html.

Diabetes UK. n.d. "Diabetes Statistics." Accessed July 3, 2021. www.diabetes.org.uk/professionals/position-statements-reports/statistics.

Dickinson, Maggie. 2019. *Feeding the Crisis: Care and Abandonment in America's Food Safety Net.* Berkeley: University of California Press. www.ucpress.edu/book/9780520307674/feeding-the-crisis.

Dinarello, Charles A. 2007. "Historical Review of Cytokines." *European Journal of Immunology* 37 (Supp. 1): S34–45. https://doi.org/10.1002/eji.200737772.

Dodd, Jodie M., Andrea R. Deussen, and Jennie Louise. 2020. "Effects of an Antenatal Dietary Intervention in Women with Obesity or Overweight on Child Outcomes at 3–5 Years of Age: LIMIT Randomised Trial Follow-Up." *International Journal of Obesity* 44 (7): 1531–35. https://doi.org/10.1038/s41366-020-0560-4.

Dodd, Jodie M., Andrew J. McPhee, Andrea R. Deussen, Jennie Louise, Lisa N. Yelland, Julie A. Owens, and Jeffrey S. Robinson. 2018. "Effects of an Antenatal Dietary Intervention in Overweight and Obese Women on 6 Month Infant Outcomes: Follow-up from the LIMIT Randomised Trial." *International Journal of Obesity* 42 (7): 1326–35. https://doi.org/10.1038/s41366-018-0019-z.

Dodd, Jodie M., Deborah Turnbull, Andrew J. McPhee, Andrea R. Deussen, Rosalie M. Grivell, Lisa N. Yelland, Caroline A. Crowther, Gary Wittert, Julie A. Owens, and Jeffrey S. Robinson. 2014. "Antenatal Lifestyle Advice for Women Who Are Overweight or Obese: LIMIT Randomised Trial." *BMJ* 348 (February). https://doi.org/10.1136/bmj.g1285.

Duden, Barbara. 1993. *Disembodying Women: Perspectives on Pregnancy and the Unborn.* Cambridge, MA: Harvard University Press.

Dumit, Joseph. 2012. *Drugs for Life: How Pharmaceutical Companies Define Our Health.* 1st ed. Durham, NC: Duke University Press.

———. 2018. "Notes toward Critical Ethnographic Scores: Anthropology and Improvisation Training in a Breached World." In *Between Matter and Method: Encounters in Anthropology and Art*, edited by G. Bakke and M. Peterson, 51–71. London: Bloomsbury Academic.

Dumit, Joseph, and Marianne de Laet. 2014 "Curves to Bodies: The Material Life of Graphs." In *Routledge Handbook on Science, Technology and Society*, edited by Daniel Lee Kleinman and Kelly Moore, 71–88. New York: Routledge.

Dupras, Charles, and Vardit Ravitsky. 2016. "Epigenetics in the Neoliberal 'Regime of Truth.'" *Hastings Center Report* 46 (1): 26–35. https://doi.org/10 .1002/hast.522.

Duster, Troy. 2005. "Race and Reification in Science." *Science* 307 (5712): 1050–51. https://doi.org/10.1126/science.1110303.

———. 2006. "Lessons from History: Why Race and Ethnicity Have Played a Major Role in Biomedical Research." *Journal of Law, Medicine & Ethics* 34 (3): 487–96. https://doi.org/10.1111/j.1748-720X.2006.00060.x.

Edu, Ugo Felicia. 2018. "When Doctors Don't Tie: Hierarchical Medicalization, Reproduction, and Sterilization in Brazil: Medicalization and Tubal Ligation in Brazil." *Medical Anthropology Quarterly* 32 (4): 556–73. https://doi.org/10 .1111/maq.12475.

———. 2019. "Aesthetics Politics: Negotiations of Black Reproduction in Brazil." *Medical Anthropology* 38 (8): 680–94. https://doi.org/10.1080/01459740.2019 .1665671.

Elattar, Sawsan, and Ande Satyanarayana. 2015. "Can Brown Fat Win the Battle against White Fat?" *Journal of Cellular Physiology* 230 (10): 2311–17. https:// doi.org/10.1002/jcp.24986.

El-Haj, Nadia Abu. 2007. "The Genetic Reinscription of Race." *Annual Review of Anthropology* 36 (1): 283–300. https://doi.org/10.1146/annurev.anthro.34 .081804.120522.

Epstein, Steven. 2009. *Inclusion: The Politics of Difference in Medical Research.* Chicago: University of Chicago Press.

Eriksson, J., T. Forsén, J. Tuomilehto, C. Osmond, and D. Barker. 2001. "Size at Birth, Childhood Growth and Obesity in Adult Life." *International Journal of Obesity and Related Metabolic Disorders* 25 (5): 735–40. https://doi.org/10 .1038/sj.ijo.0801602.

Falu, Nessette. 2019. "*Vivência Negra* : Black Lesbians Affective Experiences in Brazilian Gynecology." *Medical Anthropology* 38 (8): 695–709. https://doi.org /10.1080/01459740.2019.1666845.

Fannin, Maria, and Julie Kent. 2015. "Origin Stories from a Regional Placenta Tissue Collection." *New Genetics and Society* 34 (1): 25–51. https://doi.org/10 .1080/14636778.2014.999153.

Farquhar, Judith, and Margaret Lock, eds. 2007. *Beyond the Body Proper: Reading the Anthropology of Material Life.* 1st ed. Durham, NC: Duke University Press.

Farrar, Diane, Mark Simmonds, Susan Griffin, Ana Duarte, Debbie A. Lawlor, Mark Sculpher, Lesley Fairley, et al. 2016. "Prevalence of Gestational Diabetes in the UK and Republic of Ireland: A Systematic Review." NIHR Journals Library. www.ncbi.nlm.nih.gov/books/NBK401113/.

Feinberg, Andrew P. 2007. "Phenotypic Plasticity and the Epigenetics of Human Disease." *Nature* 447 (7143): 433–40. https://doi.org/10.1038/nature05919.

Feng, Suhua, Steven E. Jacobsen, and Wolf Reik. 2010. "Epigenetic Reprogramming in Plant and Animal Development." *Science* 330 (6004): 622–27. https://doi.org/10.1126/science.1190614.

Ferguson, James. 1994. *The Anti-Politics Machine: "Development," Depoliticization, and Bureaucratic Power in Lesotho*. Minneapolis: University of Minnesota Press.

Fessler, Pam and Joel Rose. 2019. "Trump Administration Rule Would Penalize Immigrants For Needing Benefits." NPR, August 12. www.npr.org/2019/08/12/748328652/trump-administration-rule-would-penalize-immigrants-for-using-benefits.

Fichter, Manfred Maximilian, and Norbert Quadflieg. 2016. "Mortality in Eating Disorders—Results of a Large Prospective Clinical Longitudinal Study." *International Journal of Eating Disorders* 49 (4): 391–401. https://doi.org/10.1002/eat.22501.

Fisher, Jill A. 2009. *Medical Research for Hire: The Political Economy of Pharmaceutical Clinical Trials*. Critical Issues in Health and Medicine. New Brunswick, NJ: Rutgers University Press.

———. 2020. *Adverse Events: Race, Inequality, and the Testing of New Pharmaceuticals*. New York: New York University Press.

Fisher, Jill A., and Corey A. Kalbaugh. 2011. "Challenging Assumptions about Minority Participation in US Clinical Research." *American Journal of Public Health* 101 (12): 2217–22. https://doi.org/10.2105/AJPH.2011.300279.

Fleck, Ludwik. 2008. *Genesis and Development of a Scientific Fact*. Edited by Thaddeus J. Trenn and Robert K. Merton. Translated by Fred Bradley and Thaddeus J. Trenn. Chicago: University of Chicago Press.

Flynn, Angela C., Kathryn Dalrymple, Suzanne Barr, Lucilla Poston, Louise M. Goff, Ewelina Rogozińska, Mireille N. M. van Poppel, et al. 2016. "Dietary Interventions in Overweight and Obese Pregnant Women: A Systematic Review of the Content, Delivery, and Outcomes of Randomized Controlled Trials." *Nutrition Reviews* 74 (5): 312–28. https://doi.org/10.1093/nutrit/nuw005.

Foucault, Michel. 1978. *The History of Sexuality*. Vol. I, *An Introduction*. Translated by R. Hurley. New York: Vintage.

———. 1994a. *The Birth of the Clinic: An Archaeology of Medical Perception*. New York: Vintage.

———. 1994b. *The Order of Things: An Archaeology of the Human Sciences*. Reissue ed. New York: Vintage.

———. 2003. *Society Must Be Defended: Lectures at the College de France, 1975–76*. Translated by D. Macey. Edited by M. Bertani and A. Fontana. New York: Picador.

Franklin, Sarah. 2013. *Biological Relatives: IVF, Stem Cells, and the Future of Kinship*. Experimental Futures. Durham, NC: Duke University Press.

Freccero, Carla. 2011. "Carnivorous Virility; or, Becoming-Dog." *Social Text* 29 (106)): 177–95. https://doi.org/10.1215/01642472-1210311.

Frew, P. M., D. S. Saint-Victor, M. B. Isaacs, S. Kim, G. K. Swamy, J. S. Sheffield, K. M. Edwards, T. Villafana, O. Kamagate, and K. Ault. 2014. "Recruitment and Retention of Pregnant Women into Clinical Research Trials: An Overview of Challenges, Facilitators, and Best Practices." *Clinical Infectious Diseases* 59 (Supp. 7): S400–407. https://doi.org/10.1093/cid/ciu726.

Fullwiley, Duana. 2011. *The Enculturated Gene: Sickle Cell Health Politics and Biological Difference in West Africa.* Princeton, NJ: Princeton University Press.

Gálvez, Alyshia. 2018. *Eating NAFTA.* 1st ed. Oakland: University of California Press.

Gálvez, Alyshia, Megan Carney, and Emily Yates-Doerr. 2020. "Chronic Disaster: Reimagining Noncommunicable Chronic Disease." *American Anthropologist* 122 (3): 639–40. https://doi.org/10.1111/aman.13437.

Gilmore, Ruth Wilson. 2007. *Golden Gulag: Prisons, Surplus, Crisis, and Opposition in Globalizing California.* 1st ed. Berkeley: University of California Press.

Ginsburg, Faye, and Rayna Rapp. 1991. "The Politics of Reproduction." *Annual Review of Anthropology* 20 (1): 311–43. https://doi.org/10.1146/annurev.an.20 .100191.001523.

Ginsburg, Faye D., and Rayna Rapp, eds. 1995. *Conceiving the New World Order: The Global Politics of Reproduction.* Berkeley: University of California Press.

Github. n.d. COVID-19 Data Repository by the Center for Systems Science and Engineering (CSSE) at Johns Hopkins University. https://github.com /CSSEGISandData/COVID-19.

Gluckman, Peter, and Mark Hanson. 2008. *Mismatch: The Lifestyle Diseases Timebomb.* 1st ed. Oxford: Oxford University Press.

Gluckman, Peter D., Mark A. Hanson, and Hamish G. Spencer. 2005. "Predictive Adaptive Responses and Human Evolution." *Trends in Ecology & Evolution* 20 (10): 527–33. https://doi.org/10.1016/j.tree.2005.08.001.

Gordon, Avery. 2008. *Ghostly Matters: Haunting and the Sociological Imagination.* New ed. Minneapolis: University of Minnesota Press.

Graham, H. 1996. "Smoking Prevalence among Women in the European Community 1950–1990." *Social Science & Medicine* 43 (2): 243–54. https://doi .org/10.1016/0277-9536(95)00369-x.

Greenhalgh, Susan. 2012. "Weighty Subjects: The Biopolitics of the U.S. War on Fat." *American Ethnologist* 39 (3): 471–87. https://doi.org/10.1111/j.1548-1425 .2012.01375.x.

Greenhalgh, Susan, and Megan Carney. 2014. "Bad Biocitizens? Latinos and the US 'Obesity Epidemic.'" *Human Organization* 73 (3): 267–76. https://doi.org /10.17730/humo.73.3.w53hh1t413038240.

Grieve J. F. 1974. "Prevention of Gestational Failure by High Protein Diet."
   *Journal of Reproductive Medicine* 13: 170–74.
———. 1975. "A Comment on the Relation between Protein and Weight Gain in
   Pregnancy." *Journal of Reproductive Medicine* 14: 55.
Grieve, J. K., B.M. Campbell, and F. D. Johnstone, 1978. "Dieting in Pregnancy:
   A Study of the Effect of a High Protein Low Carbohydrate Diet on Birth
   Weight on an Obstetric Population." In *Carbohydrate Metabolism in Preg-
   nancy and the Newborn*, edited by M. W. Sutherland and J. M. Stowers,
   518–33. Berlin: Springer Verlag.
Grosz, Elizabeth. 2004. *Time Travels: Feminism, Nature, Power.* Durham, NC:
   Duke University Press.
Hacking, Ian. 2001. *The Social Construction of What?* Cambridge, MA: Harvard
   University Press.
Hackman, Daniel A., Martha J. Farah, and Michael J. Meaney. 2010. "Socioeco-
   nomic Status and the Brain: Mechanistic Insights from Human and Animal
   Research." *Nature Reviews Neuroscience* 11 (9): 651–59. https://doi.org/10
   .1038/nrn2897.
Halberstam, J. Jack. 2005. *In a Queer Time and Place: Transgender Bodies,
   Subcultural Lives.* New York: New York University Press.
Hales, C. N., and D. J. P. Barker. 1992. "Type 2 (Non-Insulin-Dependent)
   Diabetes Mellitus: The Thrifty Phenotype Hypothesis." *Diabetologia* 35 (7):
   595–601. https://doi.org/10.1007/BF00400248.
Haraway, Donna J. 1988. "Situated Knowledges: The Science Question in
   Feminism and the Privilege of Partial Perspective." *Feminist Studies* 14 (3):
   575–99.
———. 1990a. *Primate Visions: Gender, Race, and Nature in the World of Modern
   Science.* 1st ed. New York: Routledge.
———. 1990b. "Reading Buchi Emecheta: Contests for Women's Experience in
   Women's Studies." *Women: A Cultural Review* 1 (3): 240–55.
———. 1997. *Modest_Witness@Second_Millennium.FemaleMan_Meets_Onco-
   Mouse: Feminism and Technoscience.* 1st ed. New York: Routledge.
Hardin, Jessica. 2018. *Faith and the Pursuit of Health: Cardiometabolic Disor-
   ders in Samoa.* New Brunswick, NJ: Rutgers University Press.
Harding, Sandra. 1986. *The Science Question in Feminism.* 1st ed. Ithaca, NY:
   Cornell University Press.
Hardy, Janet B. 2003. "The Collaborative Perinatal Project: Lessons and Legacy."
   *Annals of Epidemiology* 13 (5): 303–11. https://doi.org/10.1016/s1047-2797
   (02)00479-9.
Harney, Stefano, and Fred Moten. 2013.*The Undercommons: Fugitive Planning
   & Black Study.* Wivenhoe, UK: Minor Compositions,
Hartman, Saidiya. 2008. *Lose Your Mother: A Journey along the Atlantic Slave
   Route.* 1st ed. New York: Farrar, Straus and Giroux.

———. 2019. *Wayward Lives, Beautiful Experiments: Intimate Histories of Social Upheaval.* New York: W. W. Norton.

Harvey, David. 2005. *The New Imperialism.* Rev. ed. Oxford: Oxford University Press.

———. 2007. *A Brief History of Neoliberalism.* Oxford: Oxford University Press.

Hayek, Friedrich A. 1994. *The Road to Serfdom.* 50th anniversary ed. Chicago: University of Chicago Press.

Hayden, Cori. 2003. *When Nature Goes Public: The Making and Unmaking of Bioprospecting in Mexico.* Princeton, NJ: Princeton University Press.

Herrick, Kirsten, David I. W. Phillips, Soraya Haselden, Alistair W. Shiell, Mary Campbell-Brown, and Keith M. Godfrey. 2003. "Maternal Consumption of a High-Meat, Low-Carbohydrate Diet in Late Pregnancy: Relation to Adult Cortisol Concentrations in the Offspring." *Journal of Clinical Endocrinology and Metabolism* 88 (8): 3554–60. https://doi.org/10.1210/jc.2003-030287.

Hill, Austin Bradford. 1965. "The Environment and Disease: Association or Causation?" *Proceedings of the Royal Society of Medicine* 58 (5): 295–300.

Hoberman, John M. 2012. *Black and Blue the Origins and Consequences of Medical Racism.* Berkeley: University of California Press. http://site.ebrary.com/id/10537971.

Holliday, R. 2006. "Epigenetics: A Historical Overview." *Epigenetics* 1 (2): 76–80.

Holmes, Jonathon. 2021. "Brexit and the End of the Transition Period." The King's Fund. January 11. www.kingsfund.org.uk/publications/articles/brexit-end-of-transition-period-impact-health-care-system.

Hoover, Elizabeth. 2017. *The River Is in Us: Fighting Toxics in a Mohawk Community.* Minneapolis: University of Minnesota Press.

———. 2018. "Environmental Reproductive Justice: Intersections in an American Indian Community Impacted by Environmental Contamination." *Environmental Sociology* 4 (1): 8–21. https://doi.org/10.1080/23251042.2017.1381898.

Hsu, Chien-Ning, and You-Lin Tain. 2019. "The Good, the Bad, and the Ugly of Pregnancy Nutrients and Developmental Programming of Adult Disease." *Nutrients* 11 (4): 894. https://doi.org/10.3390/nu11040894.

Hunt, L. M., and M. S. Megyesi. 2008. "Genes, Race and Research Ethics: Who's Minding the Store?" *Journal of Medical Ethics* 34 (6): 495–500. https://doi.org/10.1136/jme.2007.021295.

Hunt, Linda M., and Mary S. Megyesi. 2008. "The Ambiguous Meanings of the Racial/Ethnic Categories Routinely Used in Human Genetics Research." *Social Science & Medicine* 66 (2): 349–61. https://doi.org/10.1016/j.socscimed.2007.08.034.

Hurd, Paul. 2010. "The Era of Epigenetic." *Briefings in Functional Genomics* 9 (5): 425–28.

Hytten, F. 1985. "Blood Volume Changes in Normal Pregnancy." *Clinics in Haematology* 14 (3): 601–12.

Inhorn, Marcia C., and Daphna Birenbaum-Carmeli. 2008. "Assisted Reproductive Technologies and Culture Change." *Annual Review of Anthropology* 37 (1): 177–96. https://doi.org/10.1146/annurev.anthro.37.081407.085230.

Institute of Medicine (IOM). 1985. *Preventing Low Birth Weight*. Washington, DC: National Academy Press.

———. 1990. *Nutrition During Pregnancy*. Washington, DC: National Academy Press.

———. 2014. *Capturing Social and Behavioral Domains in Electronic Health Records*. Washington, DC: National Academy of Science.

Institute of Medicine (IOM) and National Research Council (NRC). 2009. *Weight Gain during Pregnancy: Reexamining the Guidelines*. Washington, DC: National Academies Press.

Jablonka, Eva, and Marion Lamb. 1995. *Epigenetic Inheritance and Evolution: The Lamarckian Dimension*. Oxford: Oxford University Press.

Jablonka, Eva, and Marion J. Lamb. 2006. *Evolution in Four Dimensions: Genetic, Epigenetic, Behavioral, and Symbolic Variation in the History of Life*. 1st ed. Cambridge, MA: A Bradford Book.

Jasanoff, Sheila. 2007. *Designs on Nature: Science and Democracy in Europe and the United States*. New ed. Princeton, NJ: Princeton University Press.

Kaempf, Joseph W., Mark Tomlinson, Cindy Arduza, Shelly Anderson, Betty Campbell, Linda A. Ferguson, Mara Zabari, and Valerie T. Stewart. 2006. "Medical Staff Guidelines for Periviability Pregnancy Counseling and Medical Treatment of Extremely Premature Infants." *Pediatrics* 117 (1): 22–29. https://doi.org/10.1542/peds.2004-2547.

Kahn, Jonathan. 2012. *Race in a Bottle: The Story of BiDil and Racialized Medicine in a Post-Genomic Age*. New York: Columbia University Press.

Kelishadi, Roya, Zohreh Badiee, and Khosrow Adeli. 2007. "Cord Blood Lipid Profile and Associated Factors: Baseline Data of a Birth Cohort Study." *Paediatric and Perinatal Epidemiology* 21 (6): 518–24. https://doi.org/10.1111/j.1365-3016.2007.00870.x.

Keller, Evelyn Fox. 1995. "Gender and Science: Origin, History, and Politics." *Osiris* 10 (1): 26–38. https://doi.org/10.1086/368741.

Kenney, Martha, and Ruth Müller. 2016. "Of Rats and Women: Narratives of Motherhood in Environmental Epigenetics." *BioSocieties* 12 (1): 23–46. https://doi.org/10.1057/s41292-016-0002-7.

Kenny, Louise C., and Douglas B. Kell. 2018. "Immunological Tolerance, Pregnancy, and Preeclampsia: The Roles of Semen Microbes and the Father." *Frontiers in Medicine* 4 (January). https://doi.org/10.3389/fmed.2017.00239.

Kesselheim, A. S., and N. Shiu. 2014. "The Evolving Role of Biomarker Patents in Personalized Medicine." *Clinical Pharmacology and Therapeutics* 95 (2): 127–29. https://doi.org/10.1038/clpt.2013.185.

Keys, Ancel. 1980. *Seven Countries: A Multivariate Analysis of Death and Coronary Heart Disease*. Cambridge, MA: Harvard University Press. https://0-doi-org.pugwash.lib.warwick.ac.uk/10.4159/harvard.9780674497887.

Knight, M., K. Bunch, D. Tuffnell, H. Jayakody, J. Shakespeare, R. Kotnis, S. Kenyon, and J. J. Kurinczuk. 2018. "Saving Lives, Improving Mothers' Care—Lessons Learned to Inform Maternity Care from the UK and Ireland Confidential Enquiries into Maternal Deaths and Morbidity 2014–16." University of Oxford. www.npeu.ox.ac.uk/downloads/files/mbrrace-uk/reports/MBRRACE-UK%20Maternal%20Report%202018%20-%20Web%20Version.pdf.

Kroløkke, Charlotte. 2018. *Global Fluids: The Cultural Politics of Reproductive Waste and Value*. 1st ed. Edited by Susanne Buckley-Zistel and Ulrike Krause. London: Berghahn Books.

Kucharski, R., J. Maleszka, S. Foret, and R. Maleszka. 2008. "Nutritional Control of Reproductive Status in Honeybees via DNA Methylation." *Science* 319 (5871): 1827–30. https://doi.org/10.1126/science.1153069.

Kuhn, Thomas S. 1996. *The Structure of Scientific Revolutions*. 3rd ed. Chicago: University of Chicago Press.

Kulick, Don, and Anne Meneley. 2005. *Fat: The Anthropology of an Obsession*. New York: Tarcher.

Kuzawa, Christopher W., and Elizabeth A. Quinn. 2009. "Developmental Origins of Adult Function and Health: Evolutionary Hypotheses." *Annual Review of Anthropology* 38 (1): 131–47. https://doi.org/10.1146/annurev-anthro-091908-164350.

Lampland, Martha, and Susan Leigh Star, eds. 2008. *Standards and Their Stories: How Quantifying, Classifying, and Formalizing Practices Shape Everyday Life*. 1st ed. Ithaca, NY: Cornell University Press,

Lamoreaux, Janelle. 2016. "What If the Environment Is a Person? Lineages of Epigenetic Science in a Toxic China." *Cultural Anthropology* 31 (2): 188–214. https://doi.org/10.14506/ca31.2.03.

Landecker, Hannah. 2010. *Culturing Life: How Cells Became Technologies*. 1st ed. Cambridge, MA: Harvard University Press.

———. 2011. "Food as Exposure: Nutritional Epigenetics and the New Metabolism." *BioSocieties* 6 (2): 167–94. https://doi.org/10.1057/biosoc.2011.1.

Landecker, Hannah, and Aaron Panofsky. 2013. "From Social Structure to Gene Regulation, and Back: A Critical Introduction to Environmental Epigenetics for Sociology." *Annual Review of Sociology* 39 (1): 333–57. https://doi.org/10.1146/annurev-soc-071312-145707.

Lappé, Martine, Robbin Jeffries Hein, and Hannah Landecker. 2019. "Environmental Politics of Reproduction." *Annual Review of Anthropology* 48 (1). https://doi.org/10.1146/annurev-anthro-102218-011346.

Lappé, Martine, and Hannah Landecker. 2015. "How the Genome Got a Life Span." *New Genetics and Society* 34 (2): 152–76. https://doi.org/10.1080/14636778.2015.1034851.

Latour, Bruno. 1987. *Science in Action: How to Follow Scientists and Engineers Through Society*. Cambridge, MA: Harvard University Press.

———. 1993. *We Have Never Been Modern*. Cambridge, MA: Harvard University Press.

Lewis, Sophie. 2019. *Full Surrogacy Now: Feminism Against Family*. London: Verso.

Lock, Margaret. 2013. "The Epigenome and Nature/Nurture Reunification: A Challenge for Anthropology." *Medical Anthropology* 32 (4): 291–308. https://doi.org/10.1080/01459740.2012.746973.

Lock, Margaret M., and Vinh-Kim Nguyen. 2010. *An Anthropology of Biomedicine*. Chichester, UK: Wiley-Blackwell.

Lopez, Iris. 2008. *Matters of Choice: Puerto Rican Women's Struggle for Reproductive Freedom*. New Brunswick, NJ: Rutgers University Press.

Luxemburg, Rosa. 1913. *The Accumulation of Capital*. Translated by Agnes Schwarzschild. www.marxists.org/archive/luxemburg/1913/accumulation-capital/index.htm.

MacCormack, Carol P., and Marilyn Strathern. 1980. *Nature, Culture and Gender*. Cambridge: Cambridge University Press.

Maher, JaneMaree, Suzanne Fraser, and Jan Wright. 2010. "Framing the Mother: Childhood Obesity, Maternal Responsibility and Care." *Journal of Gender Studies* 19 (3): 233–47. https://doi.org/10.1080/09589231003696037.

Mamo, Laura, and Jennifer Ruth Fosket. 2009. "Scripting the Body: Pharmaceuticals and the (Re)Making of Menstruation." *Signs* 34 (4): 925–49. https://doi.org/10.1086/597191.

Manderson, L. 2016. "Foetal Politics and the Prevention of Chronic Disease." *Australian Feminist Studies* 31: 154–71.

Mansfield, Becky, and Julie Guthman. 2015. "Epigenetic Life: Biological Plasticity, Abnormality, and New Configurations of Race and Reproduction." *Cultural Geographies* 22 (1): 3–20. https://doi.org/10.1177/1474474014555659.

Marcus, George E. 1995. "Ethnography in/of the World System: The Emergence of Multi-Sited Ethnography." *Annual Review of Anthropology* 24: 95–117.

Marcus, George E., and Michael M. J. Fischer. 1999. *Anthropology as Cultural Critique: An Experimental Moment in the Human Sciences*. 2nd ed. Chicago: University of Chicago Press.

Martin, Courtney E. 2008. *Perfect Girls, Starving Daughters: How the Quest for Perfection Is Harming Young Women*. Reprint ed. New York: Berkley.

Martin, Emily. 1987. *The Woman in the Body: A Cultural Analysis of Reproduction*. Boston: Beacon Press.

———. 1991. "The Egg and the Sperm: How Science Has Constructed a Romance Based on Stereotypical Male-Female Roles." *Signs* 16 (3): 485–501.

Martin, Lauren Jade. 2018. "They Don't Just Take a Random Egg: Egg Selection in the United States." In *Selective Reproduction in the 21st Century*, edited by Ayo Wahlberg and Tine M. Gammeltoft, 151–70. Cham: Springer International Publishing. https://doi.org/10.1007/978-3-319-58220-7_7.

Martin, Randy. 2007. *An Empire of Indifference: American War and the Financial Logic of Risk Management*. Annotated ed. Durham, NC: Duke University Press.

Marx, Karl. 1976. *Capital*, Vol. 1, *A Critique of Political Economy*. Reprint ed. Translated by Ben Fowkes. London: Penguin Classics.

McCullough, Megan B., and Jessica Hardin, eds. 2015. *Reconstructing Obesity: The Meaning of Measures and the Measure of Meanings*. First paperback ed. Food, Nutrition, and Culture, vol. 2. New York: Berghahn Books.

M'Charek, Amade. 2013. "BEYOND FACT OR FICTION: On the Materiality of Race in Practice." *Cultural Anthropology* 28 (3): 420–42. https://doi.org/10.1111/cuan.12012.

McGoey, L., A. Wahlberg, and J. Reiss. 2011. "The Global Health Complex." *Biosocieties* 6 (1): 1–9. https://doi.org/10.1057/biosoc.2010.45.

McKee, Chris, Nora Porter, Mary Ann Hasbrouck, Jaime Ingraham, Sandra Morales, Vicki Peyton, and Heather Alderman. 2003. "Reprogenetics and Public Policy: Reflections and Recommendations." Hastings Center. www.thehastingscenter.org/wp-content/uploads/reprogenetics_and_public_policy.pdf.

McKittrick, Katherine. 2006. *Demonic Grounds: Black Women and the Cartographies of Struggle*. Minneapolis: University of Minnesota Press.

———. 2021. *Dear Science and Other Stories*. Errantries. Durham, NC: Duke University Press.

McKittrick, Katherine, ed. 2015. *Sylvia Wynter: On Being Human as Praxis*. Durham, NC: Duke University Press.

McLeish, Jenny, and Maggie Redshaw. 2019. "Maternity Experiences of Mothers with Multiple Disadvantages in England: A Qualitative Study." *Women and Birth* 32 (2): 178–84. https://doi.org/10.1016/j.wombi.2018.05.009.

McMinn, Sean, Shalina Chatlani, Ashley Lopez, Sam Whitehead, Ruth Talbot, and Austin Fast. 2021. "Across the South, COVID-19 Vaccine Sites Missing from Black and Hispanic Neighborhoods." NPR, February 5. www.npr.org/2021/02/05/962946721/across-the-south-covid-19-vaccine-sites-missing-from-black-and-hispanic-neighbor.

McPhail, Deborah, Andrea Bombak, Pamela Ward, and Jill Allison. 2016. "Wombs at Risk, Wombs as Risk: Fat Women's Experiences of Reproductive Care." *Fat Studies* 5 (2): 98–115. https://doi.org/10.1080/21604851.2016.1143754.

Mehta, Ushma J., Anna Maria Siega-Riz, and Amy H. Herring. 2011. "Effect of Body Image on Pregnancy Weight Gain." *Maternal and Child Health Journal* 15 (3): 324–32. https://doi.org/10.1007/s10995-010-0578-7.

Meloni, Maurizio. 2014. "Remaking Local Biologies in an Epigenetic Time." *Somatosphere* (blog). August 8. http://somatosphere.net/2014/remaking-local-biologies-in-an-epigenetic-time.html/.

———. 2015. "Epigenetics for the Social Sciences: Justice, Embodiment, and Inheritance in the Postgenomic Age." *New Genetics and Society* 34 (2): 125–51. https://doi.org/10.1080/14636778.2015.1034850.

———. 2016. *Political Biology—Science and Social Values in Human Heredity from Eugenics to Epigenetics.* Palgrave Macmillan. www.palgrave.com/us/book/9781137377715.

———. 2017. "Race in an Epigenetic Time: Thinking Biology in the Plural; Race in an Epigenetic Time." *British Journal of Sociology* 68 (3): 389–409. https://doi.org/10.1111/1468-4446.12248.

———. 2018. "After and Beyond the Genome: Taking Postgenomics Seriously." *Somatosphere* (blog). June 15. http://somatosphere.net/2018/after-and-beyond-the-genome.html/.

Meloni, Maurizio, and Giuseppe Testa. 2014. "Scrutinizing the Epigenetics Revolution." *BioSocieties* 9 (4): 431–56. https://doi.org/10.1057/biosoc.2014.22.

Meloni, Maurizio, John Cromby, Des Fitzgerald, and Stephanie Lloyd, eds. 2018. *The Palgrave Handbook of Biology and Society.* Palgrave Handbooks. London: Palgrave Macmillan.

Miles, R. 1989. *Racism.* London: Routledge.

Million, Dian. 2013. *Therapeutic Nations: Healing in an Age of Indigenous Human Rights.* First ed. Tucson: University of Arizona Press.

Mohanty, Chandra Talpade. 2003. *Feminism without Borders: Decolonizing Theory, Practicing Solidarity.* Durham, NC: Duke University Press.

Mol, Annemarie. 2013. "Mind Your Plate! The Ontonorms of Dutch Dieting." *Social Studies of Science* 43 (3): 379–96. https://doi.org/10.1177/0306312712456948.

Molina, Natalia. 2006. *Fit to Be Citizens? Public Health and Race in Los Angeles, 1879–1939.* American Crossroads 20. Berkeley: University of California Press.

Molina, Santiago. Forthcoming. "How Science Produces Institutions: The Practice and Politics of Genome Editing." PhD diss., University of California, Berkeley.

Montoya, Michael. 2011. *Making the Mexican Diabetic: Race, Science, and the Genetics of Inequality.* Berkeley: University of California Press.

Mora, G. Cristina. 2014. *Making Hispanics: How Activists, Bureaucrats, and Media Constructed a New American.* Chicago: University of Chicago Press.

Morales, Alberto. 2019. "Designs on Natureculture: Emigration, Esperanza, an Global Health Politics." PhD diss., University of California, Irvine.

Moran-Thomas, Amy. 2019. *Traveling with Sugar: Chronicles of a Global Epidemic*. Oakland: University of California Press.

Moses, Yolanda. 2017. "Why Do We Keep Using the Word 'Caucasian'?" *SAPIENS* (blog). February 1. www.sapiens.org/column/race/caucasian-terminology -origin/.

Mukherjee, Sudeshna, Digna R. Velez Edwards, Donna D. Baird, David A. Savitz, and Katherine E. Hartmann. 2013. "Risk of Miscarriage among Black Women and White Women in a US Prospective Cohort Study." *American Journal of Epidemiology* 177 (11): 1271–78. https://doi.org/10.1093/aje/kws393.

Mukhopadhyay, Carol. 2008. "Getting Rid of the Word 'Caucasian.'" In *Everyday AntiRacism: Getting Real about Race*, 12–16. New York: The New Press.

Mullings, L., and A. Wali. 2001. *Stress and Resilience: The Social Context of Reproduction in Central Harlem*. New York: Plenum Press.

Murphy, M. 2011. "Distributed Reproduction." In *Corpus: An Interdisciplinary Reader on Bodies and Knowledge*, edited by Monica Casper and P. Currah, 21–38. New York: Palgrave Macmillan. https://doi.org/10.1007/978-0-230 -11953-6_2.

Murphy, Michelle. 2017a. *The Economization of Life*. Durham, NC: Duke University Press.

———. 2017b. "Alterlife and Decolonial Chemical Relations." *Cultural Anthropology* 32 (4): 494–503. https://doi.org/10.14506/ca32.4.02.

Nash, Jennifer C. 2021. *Birthing Black Mothers*. Durham, NC: Duke University Press.

National Institute of Environmental Health Sciences (NIEHS). 2017. "Gene-Environment Interaction." www.niehs.nih.gov/health/topics/science/gene-env /index.cfm.

National Institute for Health and Care Excellence (NICE). 2010. *Weight Management before, during, and after Pregnancy*. United Kingdom. https://www.nice .org.uk/guidance/ph27.

National Institutes of Health (NIH). 1993. "The NIH Revitalization Act of 1993." https://grants.nih.gov/policy/inclusion/women-and-minorities/guidelines .htm#:~:text=The%20NIH%20Revitalization%20Act%20of,and %20minorities%20in%20clinical%20research.&text=The%20statute %20includes%20a%20specific,and%2C%20in%20particular%20clinical %20trials.

———. 2014. Research Portfolio Online Reporting Tools (RePORT). http://report .nih.gov.

———. 2021. "Epigenomic and Gene Expression Signatures of Racial Differences in Chronic Low Back Pain." National Institutes of Health Research Portfolio Online Reporting Tools (RePORT). https://reporter.nih.gov/search /eSvCVu3VjoiqC_uLLlGTkg/project-details/10226643.

———. n.d. "Grants and Funding." https://grants.nih.gov/grants/guide.

National Research Council. 1970. *Maternal Nutrition and the Course of Pregnancy: Report of the Committee on Maternal Nutrition, Food and Nutrition Board.* Washington, DC: National Academy of Sciences.

Nelson, Alondra. 2016. *Beacon Press: The Social Life of DNA.* Boston: Beacon Press.

Nelson, Nicole C., Kelsey Ichikawa, Julie Chung, and Momin M. Malik. 2021. "Mapping the Discursive Dimensions of the Reproducibility Crisis: A Mixed Methods Analysis." *PLOS ONE* 16(7): e0254090. https://doi.org/10.1371/journal.pone.0254090.

New York Times. 2021. "The Coronavirus in the United Kingdom: Latest Map and Case Count." Accessed July 14, 2021. www.nytimes.com/interactive/2021/world/united-kingdom-covid-cases.html.

Niewöhner, Jörg. 2011. "Epigenetics: Embedded Bodies and the Molecularisation of Biography and Milieu." *BioSocieties* 6 (3): 279–98. https://doi.org/10.1057/biosoc.2011.4.

Niven, Daniel J., T. Jared McCormick, Sharon E. Straus, Brenda R. Hemmelgarn, Lianne Jeffs, Tavish R. M. Barnes, and Henry T. Stelfox. 2018. "Reproducibility of Clinical Research in Critical Care: A Scoping Review." *BMC Medicine* 16 (February). https://doi.org/10.1186/s12916-018-1018-6.

Noble, Safiya Umoja. 2018. *Algorithms of Oppression: How Search Engines Reinforce Racism.* New York: New York University Press.

Non, Amy L. 2021. "Social Epigenomics: Are We at an Impasse?" *Epigenomics* (March 22): epi-2020-0136. https://doi.org/10.2217/epi-2020-0136.

Non, Amy L., Alexandra M. Binder, Laura D. Kubzansky, and Karin B. Michels. 2014. "Genome-Wide DNA Methylation in Neonates Exposed to Maternal Depression, Anxiety, or SSRI Medication during Pregnancy." *Epigenetics* 9 (7): 964–72. https://doi.org/10.4161/epi.28853.

Nuru-Jeter, Amani M., Elizabeth K. Michaels, Marilyn D. Thomas, Alexis N. Reeves, Roland J. Thorpe, and Thomas A. LaVeist. 2018. "Relative Roles of Race versus Socioeconomic Position in Studies of Health Inequalities: A Matter of Interpretation." *Annual Review of Public Health* 39 (1): 169–88. https://doi.org/10.1146/annurev-publhealth-040617-014230.

Nutrire CoLab. 2020. "Anthropologists Respond to *The Lancet* EAT Commission." *Bionatura: Latin American Journal of Biotechnology and Life Sciences.* https://doi.org/10.21931/RB/2020.05.01.2.

Oakley, A. 1998. "Experimentation and Social Interventions: A Forgotten but Important History." *BMJ* 317: 1239–42. https://doi.org/10.1136/bmj.317.7167.1239.

Oakley, Ann, Vicki Strange, Judith Stephenson, Simon Forrest, and Helen Monteiro. 2004. "Evaluating Processes: A Case Study of a Randomized Controlled Trial of Sex Education." *Evaluation* 10 (4): 440–62. https://doi.org/10.1177/1356389004050220.

Oken, Emily, and Matthew W. Gillman. 2003. "Fetal Origins of Obesity." *Obesity Research* 11 (4): 496–506. https://doi.org/10.1038/oby.2003.69.

Omi, M., and H. Winant. 2015. *Racial Formation in the United States*. New York: Routledge.

Palmer, Steven. 2010. *Launching Global Health: The Caribbean Odyssey of the Rockefeller Foundation*. Ann Arbor: University of Michigan Press.

Pantell, Matthew S., Rebecca J. Baer, Jacqueline M. Torres, Jennifer N. Felder, Anu Manchikanti Gomez, Brittany D. Chambers, Jessilyn Dunn, et al. 2019. "Associations between Unstable Housing, Obstetric Outcomes, and Perinatal Health Care Utilization." *American Journal of Obstetrics & Gynecology MFM* 1 (4): 100053. https://doi.org/10.1016/j.ajogmf.2019.100053.

Paradies, Yin, Jehonathan Ben, Nida Denson, Amanuel Elias, Naomi Priest, Alex Pieterse, Arpana Gupta, Margaret Kelaher, and Gilbert Gee. 2015. "Racism as a Determinant of Health: A Systematic Review and Meta-Analysis." *PloS One* 10 (9): e0138511. https://doi.org/10.1371/journal.pone.0138511.

Parisi, Melissa A., Catherine Y. Spong, Anne Zajicek, and Alan E. Guttmacher. 2011. "We Don't Know What We Don't Study: The Case for Research on Medication Effects in Pregnancy." *American Journal of Medical Genetics Part C: Seminars in Medical Genetics* 157 (3): 247–50. https://doi.org/10.1002/ajmg.c.30309.

Paul, Annie Murphy. 2010. *Origins: How the Nine Months before Birth Shape the Rest of Our Lives*. New York: Simon and Schuster.

Pendergrass, Drew, and Michelle Raji. 2017. "The Bitter Pill: Harvard and the Dark History of Birth Control." *Harvard Crimson*, September 28. www.thecrimson.com/article/2017/9/28/the-bitter-pill/.

Pentecost, Michelle, and Maurizio Meloni. 2020. "'It's Never Too Early': Preconception Care and Postgenomic Models of Life." *Frontiers in Sociology* 5. https://doi.org/10.3389/fsoc.2020.00021.

Petchesky, Rosalind. 1985. *Abortion and Women's Choice: The State, Sexuality and Reproductive Freedom*. Boston: Northeastern University Press.

Peterson, Kristin. 2014. *Speculative Markets: Drug Circuits and Derivative Life in Nigeria*. Durham, NC: Duke University Press.

Petryna, Adriana. 2009. *When Experiments Travel: Clinical Trials and the Global Search for Human Subjects*. 1st ed. Princeton, NJ: Princeton University Press.

Phelan, S. M., D. J. Burgess, M. W. Yeazel, W. L. Hellerstedt, J. M. Griffin, and M. van Ryn. 2015. "Impact of Weight Bias and Stigma on Quality of Care and Outcomes for Patients with Obesity." *Obesity Reviews: An Official Journal of the International Association for the Study of Obesity* 16 (4): 319–26. https://doi.org/10.1111/obr.12266.

Pickersgill, Martyn, Jörg Niewöhner, Ruth Müller, Paul Martin, and Sarah Cunningham-Burley. 2013. "Mapping the New Molecular Landscape: Social

Dimensions of Epigenetics." *New Genetics and Society* 32 (4): 429–47. https:// doi.org/10.1080/14636778.2013.861739.

Pico, Tommy. 2016. *IRL*. Raleigh, NC: Birds, LLC.

Polanyi, Karl. 1975. *The Great Transformation*. New York: Octagon Books.

Pollock, Anne. 2012. *Medicating Race: Heart Disease and Durable Preoccupations with Difference*. 1st ed. Durham, NC: Duke University Press Books.

Pollock, Anne, and Banu Subramaniam. 2016. "Resisting Power, Retooling Justice: Promises of Feminist Postcolonial Technosciences." *Science, Technology, & Human Values* 41 (6): 951–66. https://doi.org/10.1177 /0162243916657879.

Pollock, Mica, ed. 2008. *Everyday Antiracism: Getting Real About Race in School*. New York: The New Press.

Povinelli, Elizabeth A. 2016. *Geontologies: A Requiem to Late Liberalism*. Durham, NC: Duke University Press Books.

Prescod-Weinstein, Chanda. 2020. "Making Black Women Scientists under White Empiricism: The Racialization of Epistemology in Physics." *Signs: Journal of Women in Culture and Society* 45 (2): 421–47. https://doi.org/10 .1086/704991.

Price, Kimala. 2010. "What Is Reproductive Justice? How Women of Color Activists Are Redefining the Pro-Choice Paradigm." *Meridians* 10 (2): 42–65. https://doi.org/10.2979/meridians.2010.10.2.42.

Propper, Carol, Simon Burgess, and Denise Gossage. 2008. "Competition and Quality: Evidence From the NHS Internal Market 1991–9." *Economic Journal* 118 (525): 138–70. https://doi.org/10.1111/j.1468-0297.2007.02107.x.

Propper, Carol, Simon Burgess, and Katherine Green. 2004. "Does Competition between Hospitals Improve the Quality of Care? Hospital Death Rates and the NHS Internal Market." *Journal of Public Economics* 88 (7): 1247–72. https://doi.org/10.1016/S0047-2727(02)00216-5.

Puhl, Rebecca M., and Kelly D. Brownell. 2006. "Confronting and Coping with Weight Stigma: An Investigation of Overweight and Obese Adults." *Obesity* 14 (10): 1802–15. https://doi.org/10.1038/oby.2006.208.

Puhl, Rebecca M., and Chelsea A. Heuer. 2010. "Obesity Stigma: Important Considerations for Public Health." *American Journal of Public Health* 100 (6): 1019–28. https://doi.org/10.2105/AJPH.2009.159491.

Rafter, Nicole H. 1992. "Claims-Making and Socio-Cultural Context in the First U.S. Eugenics Campaign." *Social Problems* 39 (1): 17–34. https://doi.org/10 .2307/3096909.

Rajan, Kaushik Sunder. 2006. *Biocapital: The Constitution of Postgenomic Life*. 1st ed. Durham, NC: Duke University Press.

———, ed. 2012. *Lively Capital: Biotechnologies, Ethics, and Governance in Global Markets*. Durham, NC: Duke University Press.

Rapp, Rayna. 1999. *Testing Women, Testing the Fetus: The Social Impact of Amnio-centesis in America*. The Anthropology of Everyday Life. New York: Routledge.

———. 2011. "Reproductive Entanglements: Body, State, and Culture in the Dys/Regulation of Child-Bearing." *Social Research* 78 (3): 693–718.

———. 2018. "Epigenetics at Work." *BioSocieties* 13 (4): 780–86. https://doi.org/10.1057/s41292-017-0093-9.

Rasmussen, Kathleen M., Patrick M. Catalano, and Ann L. Yaktine. 2009. "New Guidelines for Weight Gain during Pregnancy: What Obstetrician/Gynecologists Should Know." *Current Opinion in Obstetrics & Gynecology* 21 (6): 521–26. https://doi.org/10.1097/GCO.0b013e328332d24e.

Rasmussen, Kathleen M., Ann L. Yaktine, and Institute of Medicine (U.S.), eds. 2009. *Weight Gain during Pregnancy: Reexamining the Guidelines*. Washington, DC: National Academies Press.

"Reality Check on Reproducibility." 2016. *Nature* 533 (7604): 437. https://doi.org/10.1038/533437a.

Reardon, Jenny. 2004. *Race to the Finish: Identity and Governance in an Age of Genomics*. Princeton, NJ: Princeton University Press.

———. 2017. *The Postgenomic Condition: Ethics, Justice, and Knowledge after the Genome*. 1st ed. Chicago: University of Chicago Press.

Reagan, Leslie J. 2010. *Dangerous Pregnancies: Mothers, Disabilities, and Abortion in America*. Berkeley: University of California Press.

Rehman, Waqas, Lisa M. Arfons, and Hillard M. Lazarus. 2011. "The Rise, Fall and Subsequent Triumph of Thalidomide: Lessons Learned in Drug Development." *Therapeutic Advances in Hematology* 2 (5): 291–308. https://doi.org/10.1177/2040620711413165.

Retnakaran, Ravi, Anthony J. G. Hanley, Philip W. Connelly, Mathew Sermer, and Bernard Zinman. 2006. "Ethnicity Modifies the Effect of Obesity on Insulin Resistance in Pregnancy: A Comparison of Asian, South Asian, and Caucasian Women." *Journal of Clinical Endocrinology and Metabolism* 91 (1): 93–97. https://doi.org/10.1210/jc.2005-1253.

Reverby, Susan M. 2013. *Examining Tuskegee: The Infamous Syphilis Study and Its Legacy*. Reprint ed. Chapel Hill: University of North Carolina Press.

Reynolds, Rebecca M., Keith M. Godfrey, Mary Barker, Clive Osmond, and David I. W. Phillips. 2007. "Stress Responsiveness in Adult Life: Influence of Mother's Diet in Late Pregnancy." *Journal of Clinical Endocrinology & Metabolism* 92 (6): 2208–10. https://doi.org/10.1210/jc.2007-0071.

Richardson, Sarah S. 2010. "Feminist Philosophy of Science: History, Contributions, and Challenges." *Synthese* 177 (3): 337–62. https://doi.org/10.1007/s11229-010-9791-6.

———. 2015. "Maternal Bodies in the Postgenomic Order: Gender and the Explanatory Landscape of Epigenetics." In *Postgenomics: Perspectives on*

*Biology after the Genome*, edited by Sarah S. Richardson and Hallam Stevens, 210–31. Durham, NC: Duke University Press.

———. Forthcoming. *The Maternal Imprint: The Contested Science of Maternal-Fetal Effects*. Chicago: University of Chicago Press.

Richardson, Sarah S., Cynthia R. Daniels, Matthew W. Gillman, Janet Golden, Rebecca Kukla, Christopher Kuzawa, and Janet Rich-Edwards. 2014. "Society: Don't Blame the Mothers." *Nature News* 512 (7513): 131. https://doi.org/10 .1038/512131a.

Richardson, Sarah S., and Hallam Stevens, eds. 2015. *Postgenomics: Perspectives on Biology after the Genome*. Durham, NC: Duke University Press.

Roberts, Dorothy. 1998. *Killing the Black Body: Race, Reproduction, and the Meaning of Liberty*. New York: Vintage.

———. 2009. "Race, Gender, and Genetic Technologies: A New Reproductive Dystopia?." Faculty Scholarship at Penn Law. https://scholarship.law.upenn .edu/faculty_scholarship/1421/?utm_source=scholarship.law.upenn.edu %2Ffaculty_scholarship%2F1421&utm_medium=PDF&utm_campaign =PDFCoverPages.

———. 2012. *Fatal Invention: How Science, Politics, and Big Business Re-create Race in the Twenty-First Century*. New York; London: The New Press.

Roberts, Elizabeth F. S. 2017. "What Gets Inside: Violent Entanglements and Toxic Boundaries in Mexico City." *Cultural Anthropology* 32 (4): 592–619. https://doi.org/10.14506/ca32.4.07.

Robinson, Cedric. 2000. *Black Marxism: The Making of the Black Radical Tradition*. 2nd ed. Chapel Hill: University of North Carolina Press.

Rosato, Donna. 2020. "What Your Period Tracker App Knows about You." *Consumer Reports*, January 28. www.consumerreports.org/health-privacy /what-your-period-tracker-app-knows-about-you/.

Rose, Nikolas. 2006. *The Politics of Life Itself: Biomedicine, Power, and Subjectivity in the Twenty-First Century*. Princeton, NJ: Princeton University Press.

Roseboom, T. J., J. H. van der Meulen, A. C. Ravelli, C. Osmond, D. J. Barker, and O. P. Bleker. 2001. "Effects of Prenatal Exposure to the Dutch Famine on Adult Disease in Later Life: An Overview." *Molecular and Cellular Endocrinology* 185 (1–2): 93–98. https://doi.org/10.1016/s0303-7207(01)00721-3.

Ross, Loretta, and Rickie Solinger. 2017. *Reproductive Justice: An Introduction*. 1st ed. Oakland: University of California Press.

Roy, Deboleena. 2018. *Molecular Feminisms: Biology, Becomings, and Life in the Lab*. Seattle: University of Washington Press.

Rush, David. 2001. "Maternal Nutrition and Perinatal Survival." *Nutrition Reviews* 59 (10): 315–26.Saguy, Abigail C., and Rene Almeling. 2008. "Fat in the Fire? Science, the News Media, and the 'Obesity Epidemic.'" *Sociological Forum* 23 (1): 53–83. https://doi.org/10.1111/j.1600-0838.2004.00399.x-i1.

Saguy, Abigail C., and Kevin W. Riley. 2005. "Weighing Both Sides: Morality, Mortality, and Framing Contests over Obesity." *Journal of Health Politics, Policy and Law* 30 (5): 869–923. https://doi.org/10.1215/03616878-30-5-869.

Saguy, Abigail Cope. 2013. *What's Wrong with Fat?* Oxford: Oxford University Press.

Saldaña-Tejeda, Abril. 2017. "Mitochondrial Mothers of a Fat Nation: Race, Gender and Epigenetics in Obesity Research on Mexican Mestizos." *BioSocieties* 13 (December). https://doi.org/10.1057/s41292-017-0078-8.

Salway, Sarah, Ghazala Mir, Daniel Turner, George T. H. Ellison, Lynne Carter, and Kate Gerrish. 2016. "Obstacles to 'Race Equality' in the English National Health Service: Insights from the Healthcare Commissioning Arena." *Social Science & Medicine* 152 (March): 102–10. https://doi.org/10.1016/j.socscimed.2016.01.031.

Sasser, Jade. 2014. "From Darkness into Light: Race, Population, and Environmental Advocacy." *Antipode* 46 (5): 1240–57. https://doi.org/10.1111/anti.12029.

———. 2018. *On Infertile Ground.* New York: New York University Press.

Saunders, Edward J., and Jeanne A. Saunders. 1990. "Drug Therapy in Pregnancy: The Lessons of Diethylstilbestrol, Thalidomide, and Bendectin." *Health Care for Women International* 11 (4): 423–32.

Schiebinger, Londa. 2004. *Nature's Body: Gender in the Making of Modern Science.* 2nd ed. New Brunswick, NJ: Rutgers University Press.

Schwartz, Robert S. 2001. "Racial Profiling in Medical Research." *New England Journal of Medicine* 344 (18): 1392–93. https://doi.org/10.1056/NEJM200105033441810.

Scott, Courtney, Christopher T. Andersen, Natali Valdez, Francisco Mardones, Ellen A. Nohr, Lucilla Poston, Katharina C. Quack Loetscher, and Barbara Abrams. 2014. "No Global Consensus: A Cross-Sectional Survey of Maternal Weight Policies." *BMC Pregnancy and Childbirth* 14 (1). https://doi.org/10.1186/1471-2393-14-167.

Sellers, Bakari. 2020. "What the Surgeon General Gets Wrong about African Americans and Covid-19." CNN, April 14. www.cnn.com/2020/04/14/opinions/surgeon-general-comments-covid-19-black-communities-sellers/index.html.

Shannon Sullivan. 2013. "Inheriting Racist Disparities in Health." *Critical Philosophy of Race* 1 (2): 190. https://doi.org/10.5325/critphilrace.1.2.0190.

Sharp, Gemma C., Deborah A. Lawlor, and Sarah S. Richardson. 2018. "It's the Mother! How Assumptions about the Causal Primacy of Maternal Effects Influence Research on the Developmental Origins of Health and Disease." *Social Science & Medicine* 213 (September): 20–27. https://doi.org/10.1016/j.socscimed.2018.07.035.

Shiell, Allistair, Mary Campbell-Brown, Soraya Haselden, Sian Robinson, Keith M. Godfrey, and David J. P. Barker. 2001. "High-Meat, Low-Carbohydrate Diet in

Pregnancy." *Hypertension* 38 (6): 1282–88. https://doi.org/10.1161/hy1101
.095332.

Shostak, Sara. 2013. *Exposed Science: Genes, the Environment, and the Politics of Population Health*. 1st ed. Berkeley: University of California Press.

Shostak, Sara, and Margot Moinester. 2015. "The Missing Piece of the Puzzle? Measuring the Environment in the Postgenomic Moment." In *Postgenomics: Perspectives on Biology after the Genome*, edited by Sarah S. Richardson and Hallam Stevens, 192–209. Durham, NC: Duke University Press.

Siega-Riz, Anna Maria, Lisa M. Bodnar, Naomi E. Stotland, and Jamie Stang. 2020. "The Current Understanding of Gestational Weight Gain among Women with Obesity and the Need for Future Research." *NAM Perspectives* (January). https://doi.org/10.31478/202001a.

Silliman, Jael Miriam, Marlene Gerber Fried, Loretta Ross, and Elena Gutierrez, eds. 2004. *Undivided Rights: Women of Color Organize for Reproductive Justice*. Cambridge, MA: South End Press.

Singh, Ilina. 2012. "Human Development, Nature and Nurture: Working beyond the Divide." *BioSocieties* 7 (3): 308–21. https://doi.org/10.1057/biosoc.2012.20.

Singh, Ilina, and Nikolas Rose. 2009. "Biomarkers in Psychiatry." *Nature* 460 (7252): 202–7. https://doi.org/10.1038/460202a.

Slopen, Natalie, Eric B. Loucks, Allison A. Appleton, Ichiro Kawachi, Laura D. Kubzansky, Amy L. Non, Stephen Buka, and Stephen E. Gilman. 2015. "Early Origins of Inflammation: An Examination of Prenatal and Childhood Social Adversity in a Prospective Cohort Study." *Psychoneuroendocrinology* 51 (January): 403–13. https://doi.org/10.1016/j.psyneuen.2014.10.016.

SmartStart. 2012. "Protocol." US West Coast.

Smietana, Marcin, Charis Thompson, and France Winddance Twine. 2018. "Making and Breaking Families—Reading Queer Reproductions, Stratified Reproduction and Reproductive Justice Together." *Reproductive Biomedicine & Society Online* 7 (November): 112–30. https://doi.org/10.1016/j.rbms.2018.11.001.

Spillers, Hortense J. 1987. "Mama's Baby, Papa's Maybe: An American Grammar Book." *Diacritics* 17 (2): 65–81. https://doi.org/10.2307/464747.

Stacey, Judith. 1988. "Can There Be a Feminist Ethnography?" *Women's Studies International Forum* 11 (1): 21–27. https://doi.org/10.1016/0277-5395(88)90004-0.

StandUp Trial. 2012. "Protocol." United Kingdom.

Stephenson, Judith, Nicola Heslehurst, Jennifer Hall, Danielle A. J. M. Schoena-ker, Jayne Hutchinson, Janet E. Cade, Lucilla Poston, et al. 2018. "Before the Beginning: Nutrition and Lifestyle in the Preconception Period and Its Importance for Future Health." *Lancet* 391 (10132). https://doi.org/10.1016/S0140-6736(18)30311-8.

Stout, Noelle M. 2019. *Dispossessed: How Predatory Bureaucracy Foreclosed on the American Middle Class*. California Series in Public Anthropology 44. Oakland: University of California Press.

Strathern, Marilyn. 2001. *The Gender of the Gift: Problems with Women and Problems with Society in Melanesia*. Nachdr. Studies in Melanesian Anthropology 6. Berkeley: University of California Press.

———. 2004. *Partial Connections*. Updated ed. ASAO Special Publications 3. Walnut Creek, CA: AltaMira Press.

———. 2020. *Relations: An Anthropological Account*. Durham, NC: Duke University Press.

Strimbu, Kyle, and Jorge A. Tavel. 2010. "What Are Biomarkers?" *Current Opinion in HIV and AIDS* 5 (6): 463–66. https://doi.org/10.1097/COH .0b013e32833ed177.

Strings, Sabrina. 2019. *Fearing the Black Body: The Racial Origins of Fat Phobia*. New York: New York University Press.

Sturdy, Steve, and Roger Cooter. 1998. "Science, Scientific Management, and the Transformation of Medicine in Britain c. 1870–1950." *History of Science* 36 (4): 421–66. https://doi.org/10.1177/007327539803600402.

Subramaniam, Banu. 2014. *Ghost Stories for Darwin: The Science of Variation and the Politics of Diversity*. 1st ed. Urbana: University of Illinois Press.

Subramaniam, Banu. 2019. *Holy Science: The Biopolitics of Hindu Nationalism*. Seattle: University of Washington Press.

Sugden, Karen, Eilis J. Hannon, Louise Arseneault, Daniel W. Belsky, David L. Corcoran, Helen L. Fisher, Renate M. Houts, et al. 2020. "Patterns of Reliability: Assessing the Reproducibility and Integrity of DNA Methylation Measurement." *Patterns* 1 (2): 100014. https://doi.org/10.1016/j.patter.2020 .100014.

Sullivan, Shannon. 2013. "Inheriting Racist Disparities in Health: Epigenetics and the Transgenerational Effects of White Racism." *Critical Philosophy of Race* 1 (2): 190–218. https://doi.org/10.5325/critphilrace.1.2.0190.

Sunder-Rajan, Kaushik. 2006. *Biocapital: The Constitution of Post-genomic Life*. Durham, NC: Duke University Press.

Susser, M., and Zena Stein. 1994. "Timing in Prenatal Nutrition: A Reprise of the Dutch Famine Study." *Nutrition Reviews* 52 (3): 84–94.

Sutin, A. and Terracciano, A. 2013. "Perceived Weight Discrimination and Obesity." *PLoS ONE* 8 (7):e70048.

Suzuki, Masako, Ryo Maekawa, Nicole E. Patterson, David M. Reynolds, Brent R. Calder, Sandra E. Reznik, Hye J. Heo, Francine Hughes Einstein, and John M. Greally. 2016. "Amnion as a Surrogate Tissue Reporter of the Effects of Maternal Preeclampsia on the Fetus." *Clinical Epigenetics* 8 (1): 67. https:// doi.org/10.1186/s13148-016-0234-1.

Syddall, H. E., A. Aihie Sayer, E. M. Dennison, H. J. Martin, D. J. P. Barker, and C. Cooper. 2005. "Cohort Profile: The Hertfordshire Cohort Study." *International Journal of Epidemiology* 34 (6): 1234–42. https://doi.org/10.1093/ije/dyi127.

Szyf, Moshe. 2009. "Implications of a Life-Long Dynamic Epigenome." *Epigenomics* 1 (1): 9–12. https://doi.org/10.2217/epi.09.15.

TallBear, Kim. 2013. *Native American DNA: Tribal Belonging and the False Promise of Genetic Science*. Minneapolis: University of Minnesota Press.

Thacker, Eugene. 2005. "Nomos, Nosos, and Bios." *Culture Machine* 7 (blog). https://culturemachine.net/biopolitics/nomos-nosos-and-bios/.

Thayer, Zaneta M., and Amy L. Non. 2015. "Anthropology Meets Epigenetics: Current and Future Directions: Anthropology Meets Epigenetics." *American Anthropologist* 117 (4): 722–35. https://doi.org/10.1111/aman.12351.

Thompson, Charis. 2007. *Making Parents: The Ontological Choreography of Reproductive Technologies*. 1st ed. Cambridge, MA: The MIT Press.

———. 2013. *Good Science: The Ethical Choreography of Stem Cell Research*. Cambridge, MA: The MIT Press.

Timmermans, Stefan, and Marc Berg. 2003. *The Gold Standard: The Challenge of Evidence-Based Medicine and Standards in Health Care*. Philadelphia, PA: Temple University Press.

Tomes, Nancy. 1999. *The Gospel of Germs: Men, Women, and the Microbe in American Life*. Reprint ed. Cambridge, MA: Harvard University Press.

Tompkins, Kyla Wazana. 2012. *Racial Indigestion: Eating Bodies in the 19th Century*. New York: New York University Press.

Ulijaszek, Stanley J., and Hayley Lofink. 2006. "Obesity in Biocultural Perspective." *Annual Review of Anthropology* 35 (1): 337–60. https://doi.org/10.1146/annurev.anthro.35.081705.123301.

United Kingdom Parliament. 2021. 1946 National Health Service Act. www.parliament.uk/about/living-heritage/transformingsociety/livinglearning/coll-9-health1/health-01/.

University of Edinburgh. 2005. "'Motherwell's Babies' Study May Yield up Clues for Adult Diseases." *ScienceDaily*, June 24. www.sciencedaily.com/releases/2005/06/050623001353.htm.

Valdez, Natali. 2018. "The Redistribution of Reproductive Responsibility: On the Epigenetics of 'Environment' in Prenatal Interventions." *Medical Anthropology Quarterly* 32 (3): 425–42. https://doi.org/10.1111/maq.12424.

———. 2019a. "Epigenetics and Methodological Limits." Society for Cultural Anthropology. https://culanth.org/fieldsights/epigenetics-and-methodological-limits.

———. 2019b. "Improvising Race: Clinical Trials and Racial Classification." *Medical Anthropology* 38 (8): 635–50. https://doi.org/10.1080/01459740.2019.1642887.

———. 2020. "Reproducing Whiteness: Race, Food, and Epigenetics." *American Anthropologist* 122 (3): 668–69. https://doi.org/10.1111/aman.13451.

———. Forthcoming. "Pregnant Narratives: Analyzing Prenatal Trials through a Politics of Postgenomic Reproduction Framework." *Science, Technology, and Human Values.*

Valdez, Natali, and Daisy Deomampo. 2019. "Centering Race and Racism in Reproduction." *Medical Anthropology* 38 (7): 551–59. www.tandfonline.com /doi/abs/10.1080/01459740.2019.1643855.

Valdez, Nicol. 2019. "The Elusive Dream: (Mexican) Americans and the Failed Promise of American." PhD diss., Columbia University.

van Eijk, Marieke. 2018. "The Anthropology of 'Boring' Things." *Medical Anthropology Quarterly Second Spear* (blog). July 16. http://medanthro quarterly.org/?p=446.

van Speybroeck, L. 2000. "The Organism: A Crucial Genomic Context in Molecular Epigenetics?" *Theory in Biosciences* 119 (3–4): 187–208. https://doi .org/10.1078/1431-7613-00016.

Van Wichelen, Sonja, and Jaya Keaney. Forthcoming. "Introduction: Postgenomic Reproduction." *Science, Technology and Human Values.*

Vora, Kalindi. 2015. *Life Support: Biocapital and the New History of Outsourced Labor.* 1st ed. Minneapolis: University of Minnesota Press.

Waddington, C. H. 1956. "Embryology, Epigenetics and Biogenetics." *Nature* 177 (4522): 1241. https://doi.org/10.1038/1771241a0.

Wadhwa, Pathik D., Claudia Buss, Sonja Entringer, and James M. Swanson. 2009. "Developmental Origins of Health and Disease: Brief History of the Approach and Current Focus on Epigenetic Mechanisms." *Seminars in Reproductive Medicine* 27 (5): 358–68. https://doi.org/10.1055/s-0029-1237424.

Waggoner, Miranda. 2017. *The Zero Trimester: Pre-Pregnancy Care and the Politics of Reproductive Risk.* 1st ed. Oakland: University of California Press.

Waldby, Catherine, and Robert Mitchell. 2006. *Tissue Economies: Blood, Organs, and Cell Lines in Late Capitalism.* Durham, NC: Duke University Press.

Walsh, J. M., C. A. McGowan, R. Mahony, M. E. Foley, and F. M. McAuliffe. 2012. "Low Glycaemic Index Diet in Pregnancy to Prevent Macrosomia (ROLO Study): Randomised Control Trial." *BMJ* 345: e5605–e5605. https://doi.org/10 .1136/bmj.e5605.

Ward, J., and C. Warren. 2006. *Silent Victories: The History and Practice of Public Health in Twentieth Century America.* New York: Oxford University Press.

Warin, Megan, and Natali Valdez. 2020. "#My(White)BodyMyChoice." *Thesis Eleven* (blog). July 13. https://thesiseleven.com/2020/07/14/mywhitebodymy choice/.

Warin, Megan, Tanya Zivkovic, Vivienne Moore, and Michael Davies. 2012. "Mothers as Smoking Guns: Fetal Overnutrition and the Reproduction of

Obesity." *Feminism & Psychology* 22 (3): 360–75. https://doi.org/10.1177
/0959353512445359.

Wartman, Kristin. 2013. "Bad Eating Habits Start in the Womb." *New York Times*, December 1. www.nytimes.com/2013/12/02/opinion/bad-eating-habits -start-in-the-womb.html.

Weaver, Ian C. G., Michael J. Meaney, and Moshe Szyf. 2006. "Maternal Care Effects on the Hippocampal Transcriptome and Anxiety-Mediated Behaviors in the Offspring That Are Reversible in Adulthood." *Proceedings of the National Academy of Sciences of the United States of America* 103 (9): 3480–85. https://doi.org/10.1073/pnas.0507526103.

Waterland R. A., M. Travisano, K. G. Tahiliani, M. T. Rached, and S. Mirza. 2008. "Methyl Donor Supplementation Prevents Transgenerational Amplification of Obesity." *International Journal of Obesity* 32 (9): 1373–79.

"Weight Watchers History and Philosophy" n.d. Accessed July 18, 2015. www .weightwatchers.com/about/his/history.aspx.

Weinbaum, Alys Eve. 2019. *The Afterlife of Reproductive Slavery: Biocapitalism and Black Feminism's Philosophy of History.* Durham, NC: Duke University Press.

Weir, Lorna. 2006. *Pregnancy, Risk and Biopolitics: On the Threshold of the Living Subject.* 1st ed. Abingdon, Oxon: Routledge.

White, Sue. 2017. "The Rise and Rise of Prevention Science in UK Family Welfare: Surveillance Gets under the Skin." *Families, Relationships and Societies* 6 (3): 427–45. https://doi.org/10.1332/204674315X14479283041843.

Wiles, Rose. 1994. "'I'm Not Fat, I'm Pregnant': The Impact of Pregnancy on Fat Women's Body Image." In *Women and Health: Feminist Perspectives*, edited by Sue Wilkinson and Celia Kitzinger. London: Taylor & Francis.

Wilkinson, Sue, and Celia Kitzinger, eds. 1994. *Women and Health: Feminist Perspectives.* London: Taylor & Francis.

Wing, R., M. Marcus, R. Salata, L. Epstein, S. Miaskiewicz, and H. Blair. 1991. "Effects of a Very-Low-Calorie Diet on Long-Term Glycemic Control in Obese Type 2 Diabetic Subjects." *Archives of Internal Medicine* 151 (7): 1334–40.

Wolf, Jacqueline H. 2003. "Low Breastfeeding Rates and Public Health in the United States." *American Journal of Public Health* 93 (12): 2000–2010. https://doi.org/10.2105/ajph.93.12.2000.

Wolf, Naomi. 2002. *The Beauty Myth: How Images of Beauty Are Used Against Women.* Reprint ed. New York: Harper Perennial.

World Health Organization (WHO). 2000. "Obesity: Preventing and Managing the Global Epidemic." Report of a WHO Consultation. https://pubmed.ncbi .nlm.nih.gov/11234459/.

———. 2001. *International Programme on Chemical Safety Biomarkers in Risk Assessment: Validity and Validation.* Geneva: World Health Organization. http://www.inchem.org/documents/ehc/ehc/ehc222.htm.

———, ed. 2002. *Genomics and World Health: Report of the Advisory Committee on Health Research.* Geneva: World Health Organization.

Wynter, Sylvia. 2003. "Unsettling the Coloniality of Being/Power/Truth/Freedom: Towards the Human, After Man, Its Overrepresentation—An Argument." *CR: The New Centennial Review* 3 (3): 257–337. https://doi.org/10.1353/ncr.2004.0015.

Yates-Doerr, Emily. 2015. *The Weight of Obesity: Hunger and Global Health in Postwar Guatemala.* 1st ed. Oakland: University of California Press.

———. 2020. "Reworking the Social Determinants of Health: Responding to Material-Semiotic Indeterminacy in Public Health Interventions." *Medical Anthropology Quarterly* 34 (3): 378–97. https://doi.org/10.1111/maq.12586.

———. Forthcoming. *Doing Good Science: When Fetal Development Is Global Development in Guatemala and Beyond.*

Yehuda, Rachel. 2015. "How Trauma and Resilience Cross Generations." In *On Being with Krista Tippet.* Transcript of podcast, aired July 30, updated November 9, 2017. https://onbeing.org/programs/rachel-yehuda-how-trauma-and-resilience-cross-generations-nov2017/.

Yehuda, Rachel, Nikolaos P. Daskalakis, Linda M. Bierer, Heather N. Bader, Torsten Klengel, Florian Holsboer, and Elisabeth B. Binder. 2016. "Holocaust Exposure Induced Intergenerational Effects on FKBP5 Methylation." *Biological Psychiatry* 80 (5): 372–80. https://doi.org/10.1016/j.biopsych.2015.08.005.

Yehuda, Rachel, Nikolaos P. Daskalakis, Amy Lehrner, Frank Desarnaud, Heather N. Bader, Iouri Makotkine, Janine D. Flory, Linda M. Bierer, and Michael J. Meaney. 2014. "Influences of Maternal and Paternal PTSD on Epigenetic Regulation of the Glucocorticoid Receptor Gene in Holocaust Survivor Offspring." *American Journal of Psychiatry* 171 (8): 872–80. https://doi.org/10.1176/appi.ajp.2014.13121571.

Zuboff, Shoshana. 2019. *The Age of Surveillance Capitalism: The Fight for a Human Future at the New Frontier of Power.* 1st ed. New York: PublicAffairs.

# Index

ableism, 116
adoptions, forced, 16
Almeling, Rene, 48, 213n65, 214n74, 220n92
Alzheimer's disease, 35
anorexia, 116
anthropology, 13, 22; of reproduction, 17
anti-abortion movement, 63
anticipatory regimes, 10, 185, 211n28
anti-racist interventions, 202
assisted reproductive technologies (ARTs), 213n55

Barker, David, 36–37, 46. *See also* DOHaD
biobits. *See* pregnant biobits
bioeconomies, 212n45
biological matter, as a commodity, 13
biomarkers, 33, 180–82, 189, 212n49, 233n34; development of, 14; molecular biomarkers, 181; potential value of, 14
biomedicine, 16
bioprospecting, in Mexico, 231n5
biosamples, 232n11
biotechnology, 13
Black feminists/feminism, 11, 17, 24, 183–84, 228n9
Black Marxist, as a framework, 188, 211n35. *See also* Robinson, Cedric

bodies/body parts, 231n7
body image, 116
body mass index (BMI), 2, 57, 58, 75–76, 86, 94, 118, 170, 187, 234n50; determination of as based on time, 177–78; "high" BMI, 65, 88; "low" BMI, 64; "normal" BMI, 64–65; overdependence on, 77–78
BPA, 199
breast milk production, 12
bulimia, 116

capitalism, 4, 211n35, 223n36, 231n5, 232n15; dynamic capitalisms, 11; flexibility of, 11–12; literature concerning biocapitalism, 13; racial capitalism, 11; surveillance capitalism, 12, 13, 211n35, 232n15; twenty-first century capitalism, 47–48; Western capitalism, 11. *See also* racial-surveillance biocapitalism
cardiovascular disease, 37
chromatin structure, 33
clinical research, inclusion of ethnic diversity in, 93–94
clinical research facility (CRF), 101–2
colonialism, 47; nutritional colonialism, 128
Community Care Act (1990), 63–64

FemTech, 12
feraline, 182–83
fertility, 4, 12, 213n67
fetal environment, 48–49
fetal-maternal relationship, 74–75
fetal personhood, 153, 220n94
"fetal programming," 35
Fleck, Ludwick, 223n36
food, 228n19; food as exposure and exposure
   to food, 155; reconceptualization of, 147
future health, 4, 9–11, 13, 18–19, 22, 31, 78, 83,
   100, 116–18, 1136, 166–68; investment in,
   184–87; and the future risk of disease, 6,
   84, 184–85

gender, 4, 11, 32, 43, 47, 50, 54, 149, 157, 183,
   203; cis-gender males, 193; gender bias,
   17, 18; gender binaries, 9; gender inclu-
   sive environments, 21; in postgenomic
   reproduction, 22
genetic assimilation, 33–34
genetic polymorphisms, 207
genetics, World Health Organization definition of, 224n57
genome wide studies (GWAS), 216n12
genomics, 32–33, 71; genomic reproduction,
   17. See also postgenomics
gestational age (GA), 176–77; assessment
   of, 178
gestational diabetes mellitus (GDM), 2, 86,
   158, 166–67, 184, 185–87, 193; oral glucose
   tolerance test for, 3
gestational sac, 178
gestational weight gain (GWG), 51, 57–59,
   63, 64–66, 67, 90, 115, 170, 192, 223n43;
   overdependence on, 77–78; schematic
   summary of potential determinants, consequences, and effect modifiers for, 68fig.,
   schematic summary of potential determinants, consequences, and effect modifiers
   for (modified from IOM 1990), 73fig.
Global North, 31, 39, 48, 52, 69, 112, 217n45,
   220n90; definition of, 6–7; development
   of health policies in, 43–44
glycemic control, 140
glycemic index (GI), 118–19
Grieve, J. F. Kerr, 29–30, 45–46

Haraway, Donna, 166, 214n80, 215n88,
   219n68, 230n21, 231n28, 232n7
Hartman, Saidiya, 11. See also Black
   Feminism

health/health care, 234n50; "good health,"
   225n10; infrastructures of, 7; prenatal
   health care, 210n13; privatization of health
   care, 99–100; public health campaigns,
   213n64
health justice, 201–3
Hispanic/Mexican American, as not comprehensive terms, 92–93
histone modification, 33
Human Genome Project (HGP), 32
human rights violations, 39–40
hygiene campaigns: in the Caribbean and
   Central America, 39; state-endorsed
   campaigns, 138
Hyperglycemia and Adverse Pregnancy Outcome (HAPO), 234n50

incarceration. See mass incarceration
inclusion, 93–94
indeterminacy, 137
infant adiposity, 37–38
inheritance: "inherited trauma," 197; reconceptualization of, 66; transgenerational
   inheritance, 196–98
Institute of Medicine (IOM), 57, 60, 62–63,
   64, 154; recommendations of for maternal
   weight gain and nutrition, 65, 118. See
   also Institute of Medicine (IOM), reports
   of on maternal nutrition and weight
   (1990–2009)
Institute of Medicine (IOM), reports of on
   maternal nutrition and weight (1990–2009),
   66–69; the 1990 report (main highlights
   of), 66–69, 222–23n34; the 2009 report (on
   reproducing obesity), 71–72; on epigenetic
   and obesity during pregnancy, 69–71
intellectual property, 14

Lamarck, Jean-Baptiste, 34, 216n18
Landecker, Hannah, 16, 147, 174, 230n7
Lappé, Martine, 16, 148, 171, 232n14
liberalism, 13; contemporary liberalism,
   210n21; late liberalism, 97, 210n21, 223n36
"lifestyle diseases" and the "mismatch" theory,
   46–47
lifestyle interventions, 60, 78, 138–39, 194–95;
   prenatal nutritional interventions, 140
Lock, Margaret M., 231–32n7
logics: of eugenics, 87; homophobic/
   transphobic, 16; nationalistic, 16; racist
   eugenic logics, 94; speculative and anticipatory, 10; xenophobic, 16

Founded in 1893,
UNIVERSITY OF CALIFORNIA PRESS
publishes bold, progressive books and journals
on topics in the arts, humanities, social sciences,
and natural sciences—with a focus on social
justice issues—that inspire thought and action
among readers worldwide.

The UC PRESS FOUNDATION
raises funds to uphold the press's vital role
as an independent, nonprofit publisher, and
receives philanthropic support from a wide
range of individuals and institutions—and from
committed readers like you. To learn more, visit
ucpress.edu/supportus.